GEORGE E. MAYCOCK

IEEE
Std 446-1980
(Revision of
IEEE Std 446-1974)

IEEE Recommended Practice for Emergency and Standby Power Systems for Industrial and Commercial Applications

Sponsor
Emergency and Standby Power Systems Subcommittee
of the
Power Systems Support Committee
of the
Industrial Power Systems Department
of the
IEEE Industry Applications Society

Published by
The Institute of Electrical and Electronics Engineers, Inc

Distributed in cooperation with
Wiley-Interscience, a division of John Wiley & Sons, Inc

Approved September 27, 1979
IEEE Standards Board

Joseph L. Koepfinger, *Chairman*   **Irvin N. Howell, Jr** *Vice Chairman*
Ivan G. Easton, *Secretary*

G. Y. R. Allen	Harold S. Goldberg	J. E. May
William E. Andrus	Richard J. Gowen	Donald T. Michael*
C. N. Berglund	H. Mark Grove	R. L. Pritchard
Edward Chellotti	Loering M. Johnson	F. Rosa
Edward J. Cohen	Irving Kolodny	Ralph M. Showers
Warren H. Cook	W. R. Kruesi	J. W. Skooglund
R. O. Duncan	Leon Levy	W. E. Vannah
Jay Forster		B. W. Whittington

*Member emeritus

ISBN 0-471-08031-4

Library of Congress Catalog Number 80-81019

© Copyright 1980 by

The Institute of Electrical and Electronics Engineers, Inc

*No part of this publication may be reproduced in any form,
in an electronic system or otherwise,
without the prior written permission of the publisher.*

Foreword

(This Foreword is not a part of IEEE Std 446-1980, IEEE Recommended Practice for Emergency and Standby Power Systems for Industrial and Commercial Applications.)

In 1968 the Industrial and Commercial Power Systems (I&CPS) Committee within the Industry and General Applications Group (IGA) of the Institute of Electrical and Electronics Engineers (IEEE) recognized that a need existed for a publication which would provide guidance to industrial users and suppliers of emergency and standby power systems.

The nature of electric power failures, interruptions, and their duration cover the range in time from microseconds to days. Voltage excursions occur within the range from 20 times normal or more to a complete absence of voltage. Frequency excursions vary as widely in the form of harmonics to direct current.

These variables occur due to a multitude of conditions, both in the power system ahead of the point of the user's service entrance and following the service entrance within the user's area of distribution.

Such causes as lightning, automobiles striking power poles, ice storms, tornadoes, switching to alternate lines, and equipment failure are but a few of the causes of variables in the electric power supply ahead of the service entrance.

Within the user's area of distribution are such items as shorted and open circuits, undersized feeders, equipment failures, operator errors, temporary overloads, single phasing unbalanced feeders, fire, switching, and many other generators of variables.

In the past the demand for reliable electric power was less critical. If power was completely interrupted too often, another source was found. If voltage varied enough to cause a problem, a regulator or a larger conductor was installed.

As processes, controls, and instrumentation became more sophisticated and interlocked, the demand developed to shorten the length of outages. Increased safety standards for people required emergency and exit lighting. Many factories added medical facilities which needed reliable electric power.

With the advent of solid-state electronics and computers, the need for continuous, reliable, high-quality electric power became critical. Many installations required uninterruptible power, virtually free of frequency excursions and voltage dips, surges, and transients.

In 1969 a working group was established under the Industrial Plants Power Systems Subcommittee of the I&CPS to collect data and produce a publication entitled "Emergency Power Systems for Industrial Plants." Later that year the scope of the work was enlarged to include standby power since in meeting various needs the two systems were often found to be intertwined or one system served multiple purposes.

As the work progressed it became apparent that industrial and commercial needs contained more similarities than differences. Systems available to supply the required power for industry were found applicable to both fields. Once again the scope of the work was expanded by including commercial requirements. The working group was changed to the status of a subcommittee under the I&CPS, and the proposed publication was directed toward establishing recommended practices.

This first revision includes new and updated material. The sections on maintenance, protection, grounding, and industry applications have been added. Information provided by the Instrument Society of America's Committee on Emergency Power Supplies (SP54) has been included.

At the time it recommended these practices, the subcommittee had the following members and contributors:

Donald W. McWilliams, *Chairman*

G. M. Bauer	J. Ripley
A. W. Carden	R. W. Rosko
R. Castenschiold	M. A. Smith
K. Chen	L.H. Soderholm
G. Hensel	J. C. Solt
T. Key	J. Stewart
A.C. Lordi	C. Teague
R. Loewe	W. Timmler
D. T. Michael	R. West
P. O'Donnell	D. Zipse

Contents

SECTION	PAGE
1. Scope	17
2. Definitions	19
3. General Need Guidelines	21
3.1 Lighting	42
3.1.1 Introduction	42
3.1.2 Lighting for Evacuation Purposes	42
3.1.3 Perimeter and Security Lighting	42
3.1.4 Warning Lights	42
3.1.5 Health-Care Facilities	42
3.1.6 Standby Lighting for Equipment Repair	42
3.1.7 Lighting for Production	42
3.1.8 Lighting to Reduce Hazards to Machine Operators	43
3.1.9 Supplemental Lighting for High-Voltage Discharge Systems	43
3.1.10 Codes, Rules, and Regulations	43
3.1.11 Recommended Systems	43
3.2 Startup Power	43
3.2.1 Introduction	43
3.2.2 Example of System Utilizing Startup Power	44
3.2.3 Lighting	45
3.2.4 Engine-Driven Generators	45
3.2.5 Battery Systems	45
3.2.6 Other Systems	45
3.2.7 System Justification	45
3.3 Transportation	45
3.3.1 Introduction	45
3.3.2 Elevators	45
3.3.3 Conveyors and Escalators	47
3.3.4 Other Transportation Systems	47
3.4 Mechanical Utility Systems	47
3.4.1 Introduction	47
3.4.2 Typical Utility Systems for which Reliable Power May Be Necessary	47
3.4.3 Orderly Shutdown of Mechanical Utility Systems	48

SECTION			PAGE
	3.4.4	Alternates to Orderly Shutdown	48
3.5	Heating		48
	3.5.1	Maintaining Steam Production	48
	3.5.2	Process Heating	49
	3.5.3	Building Heating	50
3.6	Refrigeration		50
	3.6.1	Requirements of Selected Refrigeration Applications	50
	3.6.2	Refrigeration to Reduce Hazards	51
	3.6.3	Typical System to Maintain Refrigeration	51
3.7	Production		51
	3.7.1	Justification for Maintaining Production in an Industrial Facility	51
	3.7.2	Equations for Determining Cost of Power Interruptions	52
	3.7.3	Commercial Buildings	52
	3.7.4	Additional Losses Due to Power Interruptions	52
	3.7.5	Determining Likelihood of Power Failures	52
	3.7.6	Factors that Increase Likelihood of Power Failures	53
	3.7.7	Power Reserves	53
	3.7.8	Examples of Standby Power Applications for Production	53
	3.7.9	Types of Systems to Consider	53
3.8	Space Conditioning		54
	3.8.1	Definition	54
	3.8.2	Description	54
	3.8.3	Codes and Standards	54
	3.8.4	Application Considerations	54
	3.8.5	Examples of Space Conditioning Where Auxiliary Power May Be Justified	55
	3.8.6	Typical Auxiliary Power Systems	56
3.9	Fire Protection		56
	3.9.1	Codes, Rules, and Regulations	56
	3.9.2	Arson	56
	3.9.3	Typical Needs	56
	3.9.4	Application Considerations	57
	3.9.5	Feeder Routing to Fire Protection Equipment	57
3.10	Data Processing		57
	3.10.1	Classification of Systems	57
	3.10.2	Needs of Data Processing Equipment from a User's Viewpoint	59
	3.10.3	Power Requirements for Data Processing Equipment	60
	3.10.4	Influence Factors of Data Processing Systems on Incoming or Supplementary Independent Power Sources	65
	3.10.5	Power Quality Improvement Techniques	69
3.11	Life Support and Life Safety System		74
	3.11.1	Introduction	74

SECTION		PAGE
	3.11.2 Health-Care Facilities	74
	3.11.3 Other Critical Life Systems	78
3.12	Communication Systems	78
	3.12.1 Description	78
	3.12.2 Commonly Used Auxiliary Power Systems	78
	3.12.3 Evaluating Need for Auxiliary Power System	79
3.13	Signal Circuits	80
	3.13.1 Description	80
	3.13.2 Signal Circuits in Health-Care Facilities	80
	3.13.3 Signal Circuits in Industrial and Commercial Buildings	80
	3.13.4 Types of Auxiliary Power Systems	80
3.14	Standards References	80
3.15	References and Bibliography	81
	3.15.1 References	81
	3.15.2 Bibliography	81
4. Systems and Hardware		83
4.1	Guidelines for Use	83
4.2	Engine-Driven Generators	85
	4.2.1 Introduction	85
	4.2.2 Diesel-Engine Generators	85
	4.2.3 Gasoline-Engine Generators	85
	4.2.4 Gas-Engine Generators	85
	4.2.5 Derating Requirements	85
	4.2.6 Multiple-Engine Generator Set Systems	86
	4.2.7 Construction and Controls	87
	4.2.8 Typical Engine Generator Systems	87
	4.2.9 Special Considerations	90
	4.2.10 Engine Generator Set Rating	90
	4.2.11 Motor-Starting Considerations	90
	4.2.12 Load Transient Considerations	91
	4.2.13 Manual Systems	91
	4.2.14 Automatic Systems	91
	4.2.15 Automatic Transfer Devices	91
	4.2.16 Engine Generator Set Reliability	91
	4.2.17 Air Supply and Exhaust	92
	4.2.18 Noise Reduction	92
	4.2.19 Fuel Systems	92
	4.2.20 Governors and Regulation	92
	4.2.21 Starting Methods	93
	4.2.22 Lighting and Battery Charging	93
	4.2.23 Advantages and Disadvantages of Diesel-Driven Generators	93
	4.2.24 Additional Information	93
4.3	Multiple Utility Services	94
	4.3.1 Introduction	94
	4.3.2 Closed-Transition Transfer	94

SECTION			PAGE
	4.3.3	Utility Services Separation	94
	4.3.4	Simple Automatic Transfer Schemes	94
	4.3.5	Overcurrent Protection	95
	4.3.6	Transfer Device Ratings and Accessories	95
	4.3.7	Voltage Tolerances	98
	4.3.8	Transferring Motor Loads	99
	4.3.9	Operation of a Typical System	101
	4.3.10	Conclusion	104
4.4	Turbine-Driven Generators		104
	4.4.1	Introduction	104
	4.4.2	Steam Turbine Generators	104
	4.4.3	Gas and Oil Turbine Generators	104
	4.4.4	Advantages and Disadvantages	108
4.5	Mobile Equipment		108
	4.5.1	Introduction	108
	4.5.2	Special Requirements	108
	4.5.3	Special Precautions	111
	4.5.4	Maintenance	112
	4.5.5	Application	112
	4.5.6	Rental	112
	4.5.7	Fuel Systems	113
	4.5.8	Agricultural Applications	113
4.6	Mechanical Stored Energy Systems		114
	4.6.1	Introduction	114
	4.6.2	Typical System Types	114
	4.6.3	Buffer Performance	119
4.7	Battery Systems		121
	4.7.1	Introduction	121
	4.7.2	Application Information	121
	4.7.3	Recharge/Equalize Charging	121
	4.7.4	Battery Sizing	122
	4.7.5	Unit Lighting Equipment	124
	4.7.6	Central Battery Lighting Systems	125
	4.7.7	Factors to Consider When Selecting Emergency Lighting Systems	125
	4.7.8	Multiple Sources Used for Normal Lighting	126
4.8	Battery/Inverter Systems		126
	4.8.1	Introduction	126
	4.8.2	Battery/Inverter Supply Used as Standby Source	126
	4.8.3	Nonredundant Uninterruptible Power Supply	127
	4.8.4	Redundant Uninterruptible Power Supply	132
	4.8.5	Nonredundant Uninterruptible Power Supply with Static Bypass Switch	133
	4.8.6	Parallel Redundant Uninterruptible Power Supply	133
	4.8.7	Cold Standby Redundant Power Supply	133

SECTION			PAGE
	4.8.8	Parallel Nonredundant Uninterruptible Power Supply with Static Bypass Switch	135
	4.8.9	Combination Static Inverter and Rotating Uninterruptible Power Supply	135
	4.8.10	Combination Static Inverter and Engine Generator Uninterruptible Power Supply	135
	4.8.11	UPS Battery Selection	137
	4.8.12	Cost to Operate a Nonredundant UPS	138
	4.8.13	Special Precautions	139
4.9	Standards References		139
4.10	References and Bibliography		140
	4.10.1	References	140
	4.10.2	Bibliography	140
5. Maintenance			143
5.1	Introduction		143
5.2	Internal Combustion Engines		144
	5.2.1	Typical Maintenance Schedule	144
5.3	Gas Turbines		145
	5.3.1	General	145
	5.3.2	Operating Factors Affecting Maintenance	145
	5.3.3	Typical Maintenance Schedule	145
5.4	Generators		146
5.5	Static Uninterruptible Power Supplies		147
5.6	Batteries		148
	5.6.1	All-Liquid Electrolyte Batteries	148
	5.6.2	Lead-Acid Batteries	148
	5.6.3	Nickel-Iron-Alkaline Batteries	148
	5.6.4	Nickel-Cadmium Batteries	148
	5.6.5	Maintenance Interval	149
5.7	Automatic Transfer Switches		149
5.8	Conclusions		149
5.9	References		149
6. Protection			151
6.1	Introduction		151
6.2	Short-Circuit Current Considerations		151
6.3	Transfer Devices		152
	6.3.1	Codes and Standards	152
	6.3.2	Withstand Ratings	153
	6.3.3	Significance of X/R Ratio	153
	6.3.4	Transfer Switch Dielectric Strength	154
	6.3.5	Protection with Circuit Breakers	154
	6.3.6	Protection with Fuses	157
	6.3.7	Static Transfer Switches	158
6.4	Generator Protection		159
	6.4.1	Codes and Standards	159

SECTION			PAGE
	6.4.2	Main Winding Protection	159
	6.4.3	Rotor and Excitation System Protection	160
	6.4.4	Parallel Operation	160
6.5	Prime Mover Protection		160
	6.5.1	General Requirements	160
	6.5.2	Equipment Malfunction Protection	161
	6.5.3	Fuel System Protection	162
6.6	Electric Utility Power Supply		163
6.7	Uninterruptible Power Supply (UPS)		163
	6.7.1	Battery Protection	163
	6.7.2	Battery Charger Protection	164
	6.7.3	Inverter Protection	164
6.8	Equipment Physical Protection		165
6.9	Grounding		165
6.10	Conclusions		165
6.11	Standards References		165
6.12	Bibliography		166
7. Grounding			167
7.1	Introduction		167
	7.1.1	General	167
	7.1.2	Circuit Protective Equipment	167
	7.1.3	System and Equipment Grounding	167
7.2	System and Equipment Grounding Functions		168
	7.2.1	General	168
	7.2.2	System Grounding Functions	168
	7.2.3	Equipment Grounding Functions	169
7.3	Supplemental Equipment Bonding		170
7.4	Objectionable Current Through Grounding Conductors		171
7.5	System Grounding Requirements		172
7.6	Types of Equipment Grounding Conductors		173
7.7	Grounding for Separately Derived and Service-Supplied Systems		173
7.8	Grounding Arrangements for Emergency and Standby Power Systems		174
7.9	Systems with a Grounded Circuit Conductor		175
	7.9.1	Solidly Interconnected Multiple-Grounded Neutral	175
	7.9.2	Neutral Conductor Transferred by Transfer Means	177
	7.9.3	Neutral Conductor Isolated by a Transformer	181
	7.9.4	Solidly Interconnected Neutral Conductor Grounded at Service Equipment Only	183
7.10	Systems Without a Grounded Circuit Conductor		185
	7.10.1	Solidly Grounded Service	186
	7.10.2	High Resistance Grounded Service	188
7.11	Mobile Engine Generator Sets		189
7.12	Standards References		192

SECTION	PAGE
7.13 Bibliography	192
8. Industry Applications	193
8.1 Introduction	193
8.2 Glass Industry	193
8.2.1 Introduction	193
8.2.2 Applications	193
8.3 Rural Electric Power	194
8.3.1 Introduction	194
8.3.2 Full-Load Systems	194
8.3.3 Part-Load Systems	194
8.3.4 Transfer Switch	194
8.3.5 Basic Standby Generator Types	194
8.3.6 Sizing the Alternator	195
8.3.7 Installation	195
8.3.8 Maintenance	195
8.3.9 Accessory Equipment	195
8.4 Cement Industry	195
8.4.1 Introduction	195
8.4.2 Immediate Need for Power	197
8.4.3 Periodic Need for Power	197
8.4.4 Sustained Short Time Need for Power	199
8.4.5 Sample Specifications for Emergency Engine Generator Set	199
8.4.6 Sample Specifications for Uninterruptible Power Supply	201
9. Case Histories	203
9.1 Introduction	203
9.2 Determining User Needs	203
9.2.1 Earthquake Damage	203
9.2.2 Computer Power	203
9.3 Failures Due to System Design Deficiencies	203
9.3.1 Improperly Located Transfer Switch	203
9.3.2 Insufficient Compressed Air for Starting	203
9.4 Failures Caused by Lack of Maintenance	204
9.4.1 Air Base Immobilized	204
9.4.2 Drawbridge Inoperable	204
9.4.3 Mechanical Stored Energy System	204
9.5 Misapplications of Emergency or Standby Power Systems	204
9.5.1 Computer Power	204
9.6 Successful Operations of Emergency or Standby Power Systems	204
9.6.1 Bridges	204
9.6.2 UPS Survives Earthquake	204

SECTION	PAGE
Acknowledgments	207
Index	206

FIGURES		PAGE
Fig 1	Average Number of Thunderstorm Days per Year	22
Fig 2	Approximate Density of Tornadoes	22
Fig 3	Elevator Emergency Transfer System	46
Fig 4	Typical Design Goals of Power Conscious Computer Manufacturers	62
Fig 5	Typical Hospital Wiring Arrangement	76
Fig 6	Typical Engine-Driven Generator; Diesel, Gasoline, or Gas Fueled	86
Fig 7	Two Engine Generator Sets Operating in Parallel	88
Fig 8	Peaking Power Control System	88
Fig 9	Three-Source Priority Load Selection System	89
Fig 10	Combination On-Site Power and Emergency Transfer System	89
Fig 11	Dual-Engine-Generator Standby System	90
Fig 12	Two-Utility-Source System Using One Automatic Transfer Switch	94
Fig 13	Two-Utility-Source System where any Two Circuit Breakers Can be Closed	95
Fig 14	Two Utility Sources Combined with an Engine Generator Set to Provide Varying Degrees of Emergency Power	96
Fig 15	Modular Type Automatic Transfer Switch Suitable for all Classes of Load	98
Fig 16	Inphase Motor Load Transfer	99
Fig 17	Motor Load Disconnect Circuit	100
Fig 18	Neutral Off Position	101
Fig 19	Closed Transition Transfer	102
Fig 20	Typical System Supplying Power To Manufacturing Plant	103
Fig 21	Emergency and Standby Power System Using Steam-Turbine and Dual Utility Supply	105
Fig 22	Typical Gas-Turbine Generator and Riser Diagram of Auxiliary Power System	106
Fig 23	Modular Packaged Gas-Turbine Generator Set Mounted on Trailer	107
Fig 24	Typical Performance Correction Factor for Altitude	108
Fig 25	Typical Trailer Mounted Model 15 to 45 kW Capacity	110
Fig 26	Typical 2800 kW Mobile Turbine-Driven Generator	111
Fig 27	Simple Inertia-Driven "Ride Through" System	115
Fig 28	Battery-Supported Inertia System	116

FIGURES	PAGE
Fig 29 Diagram of Battery Supported Inertia System	116
Fig 30 Constant Frequency Inertia System	117
Fig 31 Rotating Flywheel No-Break System	118
Fig 32 Engine-Generator Supported Battery Inertia System	119
Fig 33 Steam-Turbine-Driven Emergency Power System	120
Fig 34 Derating Curves for Battery Chargers Due to Altitude and Temperature	123
Fig 35 Typical Battery Unit	124
Fig 36 Short-Interruption Standby System	127
Fig 37 Oscillogram of Output Voltage of System of Fig 36 During Transfer	127
Fig 38 Nonredundant Uninterruptible Power Supply	128
Fig 39 Oscillogram of System of Fig 38 with Power Line Failure	128
Fig 40 Redundant Uninterruptible Power Supply	129
Fig 41 Oscillogram of Output Voltage of System of Fig 40 upon Inverter Failure	129
Fig 42 Uninterruptible Power Supply with Static Bypass	132
Fig 43 Oscillogram of Static Switch of System of Fig 42 Load Voltage	132
Fig 44 Parallel-Supplied, Parallel Redundant Uninterruptible Power Supply	134
Fig 45 Cold Standby Redundant Uninterruptible Power Supply	134
Fig 46 Parallel-Supplied Nonredundant Uninterruptible Power Supply	135
Fig 47 Combination Static, Battery, and Rotating Uninterruptible Power Supply	136
Fig 48 Load Frequency and Voltage Stability of a UPS Under Varying Input Conditions	136
Fig 49 Typical Block Diagram of Combination Uninterruptible Power Supply and Engine Generator	137
Fig 50 Typical Battery, Rated 60 V, 1800 A · h, at 8h Discharge Rate	138
Fig 51 Fault Current Decay	152
Fig 52 One-Line Diagram	156
Fig 53 Coordination of Protective Devices on Fig 52	156
Fig 54 Emergency Power System with All Fuse Protection	157
Fig 55 Coordination Chart of Emergency Power System with All Fuse Protection	158
Fig 56 Battery Charger Regulation Curve	164
Fig 57 System and Equipment Grounding for Solidly Grounded, Service Supplied System	168
Fig 58 System and Equipment Grounding for Solidly Grounded, Separately Derived System	169

FIGURES	PAGE
Fig 59 Supplemental Equipment Bonding for Separately Derived System	170
Fig 60 Stray Neutral Current Due to Multiple Grounding of Grounded Service Conductor	171
Fig 61 Stray Neutral Current Due to Unintentional Grounding of a Grounded Circuit Conductor	172
Fig 62 System and Equipment Grounding for Separately Derived and Service Supplied Systems	174
Fig 63 Solidly Interconnected Neutral Conductor Grounded at Service Equipment and at Source of Alternate Power Supply	176
Fig 64 Stray Neutral Currents Due to Grounding the Neutral Conductor at Two Locations	177
Fig 65 Ground-Fault Current Return Paths to Normal Supply, Neutral Grounded at Two Locations	178
Fig 66 Ground-Fault Current Return Paths to Alternate Supply, Neutral Grounded at Two Locations	179
Fig 67 Transferred Neutral Conductor Grounded at Service Equipment and At Source of Alternate Supply	180
Fig 68 Stray Neutral Currents Due to Unintentionally Grounded Neutral Conductor	181
Fig 69 Transferred Neutral Conductor Grounded at Service Equipment and at Switchgear for Two On-Site Generators Connected in Parallel	182
Fig 70 Solidly Grounded Neutral Conductor for Transferred Load Isolated by Transformer	183
Fig 71 Ground-Fault Current Return Paths for Transferred Load Isolated by Transformer	184
Fig 72 Solidly Interconnected Neutral Conductor Grounded at Service Equipment Only	185
Fig 73 Ground-Fault Current Return Path to Alternate Supply, Neutral Conductor Grounded at Service Equipment Only	186
Fig 74 Three Pole Transfer Switch for Transfer to an Alternate Power Supply Without a Grounded Circuit Conductor	187
Fig 75 Interlocked Circuit Breakers for Transfer to an Alternate Power Supply Without a Grounded Circuit Conductor	188
Fig 76 High Resistance Grounded Alternate Power Supply Without a Grounded Circuit Conductor	189
Fig 77 High Resistance Grounded Systems for Normal Service and Alternate Supply	190
Fig 78 Ground-Fault Current Return Path to High	

FIGURES	PAGE
Resistance Grounded Alternate Supply	191
Fig 79 Flow Diagram for Typical Cement Plant	196
Fig 80 Diesel Generator Emergency Power for Cement Plant	198

TABLES		PAGE
Table 1	Codes for Emergency Power by States and Major Cities	29
Table 2	Condensed General Criteria for Preliminary Consideration	33
Table 3	Typical Emergency and Standby Lighting Recommendations	44
Table 4	Systems for Continued Steam Production	49
Table 5	Example of Recorded Power Failures	53
Table 6	Example of Recorded Short-Term Dips	53
Table 7	Typical Range of Input Power Quality and Load Parameters of Major Computer Manufacturers	60
Table 8	Performance of Power Conditioning Equipment	70
Table 9	Summary of Typical Power-Line Disturbances	72
Table 10	Relative Effectiveness of Power Enhancement Projects in Eliminating or Moderating Power Disturbances (US Navy)	73
Table 11	Sensitive Hospital Loads	77
Table 12	Typical Ratings of Engine-Driven Generators Approximate 1978 Prices	87
Table 13	Typical Rental/Purchase Costs for Gasoline-Powered Units (8 h Cycle, 1978 Data)	113
Table 14	Typical Rental/Purchase Costs for Diesel-Powered Units (8 h Cycle, 1978 Data)	114
Table 15	Typical Rental/Purchase Costs for Diesel-Powered Units (12 h Cycle)	114
Table 16	Power Buffer Performance of Typical Mechanical Stored Energy Systems	120
Table 17	Number of Cells for Desired Voltage	122
Table 18	General Differences for Various Battery Types	124
Table 19	Typical Nonredundant 3ϕ UPS Performance Specifications	130
Table 20	1979 Budgetary Estimates for Uninstalled Nonredundant UPS Systems Exhibiting Typical Performance Parameters of Table 19	139
Table 21	Example Withstand Current Ratings for Automatic Transfer Switches	154
Table 22	Molded-Case and Power Circuit Breaker Interrupting Requirements Per UL Std 489 and IEEE Std 20-1973	155

TABLES		PAGE
Table 23	Automatic Transfer Switch Withstand Requirements Per UL Std 1008	155
Table 24	Fuse Interrupting Test Requirements Per UL Stds 198B, 198.2, 198.3, 198.4, 198H, and ANSI C97.1-1972	155

1. Scope

This publication presents the recommended engineering principles and practices for the selection and application of emergency and standby power systems. Industrial and commercial user needs and recommended solutions to those needs are detailed.

The effects of electrical disturbances are a threefold problem, requiring close cooperation among the industrial or commercial user, the electric utility, and the equipment manufacturer. This work has been written from the user's viewpoint and directed toward his need. But this is not possible to present without also detailing the vital parts played by the other two members. Each is a leg of a triangle necessary to complete the whole figure.

Those responsible for reliable plant operation will find general operational needs itemized with the recommended technical parameters of the electric power supply required.

Commercial facility designers, operators, and owners will find operational needs listed, with references to mandatory requirements of laws, regulations, codes, and standards.

Electrical utility companies can utilize the "General Need Guidelines" in identifying the continually changing user requirements for continuous power within close tolerances. From this knowledge further assistance to the user can be provided to meet his needs.

Equipment manufacturers and system engineers will find information in the "General Need" section which will assist them in designing the necessary hardware and systems to fulfill the requirements.

Specific technical and economic information is included on available equipment and systems with recommendations to meet the requirements for various types of installations.

Within the content, the user will find the answers to the following questions:

(1) Is an emergency or standby power system, or both, needed, and what will it accomplish?

(2) What types of systems are available and which can best meet the needs?

(3) What is the estimated purchased and installed price range of the various systems?

(4) What are the operating and

maintenance requirements for maintaining system reliability?

(5) Where can additional information be obtained?

Although some needs and solutions overlap, the following fields have been excluded: aircraft, government, marine, military, mining, and utility.

2. Definitions

All definitions are listed in ANSI/IEEE Std 100-1977, Dictionary of Electrical and Electronics Terms.

availability. The fraction of time that a system is actually capable of performing its mission.

commercial power. Power furnished by an electric power utility company; when available, it is usually the prime power source.

computer. (1) A machine for carrying out calculations. (2) By extension, a machine for carrying out specified transformations on information.

controller, automatic (process control). A device that operates automatically to regulate a controlled variable in response to a command and a feedback signal.

data processing. Pertaining to any operation or combination of operations on data.

data processor. Any device capable of performing operations on data, for example, a desk calculator, a tape recorder, an analog computer, a digital computer.

dropout voltage (or current). The voltage (or current) at which a magnetically operated device will release to its de-energized position.

emergency power system. An independent reserve source of electric energy which, upon failure or outage of the normal source, automatically provides reliable electric power within a specified time to critical devices and equipment whose failure to operate satisfactorily would jeopardize the health and safety of personnel or result in damage to property.

firm power. Power intended to be always available even under emergency conditions.

forced outage. A power outage that results from failure of a system component requiring that it be taken out of service immediately, either automatically or by manual switching operations, or an outage caused by improper operation of equipment due to human error.

frequency droop. The absolute change in frequency between steady state no load and steady state full load.

frequency regulation. The percentage change in frequency from steady state no load to steady state full load, which is a function of the engine and governing system:

$$\% R = \frac{F_{n_1} - F_{f_1}}{F_{n_1}} \times 100$$

harmonic content. A measure of the presence of harmonics in a voltage or current waveform expressed as a percentage of the amplitude of the fundamental frequency at each harmonic frequency. The total harmonic content is expressed as the square root of the sum of the squares of each of the harmonic amplitudes (expressed as a percentage of the fundamental).

load shedding. The process of deliberately removing preselected loads from a power system in response to an abnormal condition in order to maintain the integrity of the system.

off line operation. Pertaining to computer systems not under direct control of the central processing unit.

on line operation. Pertaining to equipment or devices under direct control of the central processing unit.

power failure. Any variation in electric power supply which causes unacceptable performance of the user's equipment.

power outage. Complete absence of power at the point of use.

prime power. That source of supply of electric energy utilized by the user which is normally available continuously day and night, usually supplied by an electrical utility company but sometimes by the owner generation.

real time (processing). Pertaining to the actual time during which a physical process transpires or pertaining to the performance of a computation during the actual time that the related physical process transpired in order that results of the computation can be used in guiding the physical process.

redundancy. Duplication of elements in a system or installation for the purpose of enhancing the reliability or continuity of operation of the system or installation.

scheduled outage. A power outage that results when a component is deliberately taken out of service at a selected time, usually for purposes of construction, preventive maintenance, or repair.

standby power system. An independent reserve source of electric energy which, upon failure or outage of the normal source, provides electric power of acceptable quality and quantity so that the user's facilities may continue in satisfactory operation.

transient. That part of the change in a variable that disappears during transition from one steady-state operating condition to another.

uninterruptible power supply (UPS). A system designed to provide power, without delay or transients, during any period when the normal power supply is incapable of performing acceptably.

utility power (see commercial power).

3. General Need Guidelines

While all who use electric power desire perfect frequency, voltage stability, and reliability at all times, this cannot be realized in practice. Within a complex facility, the requirements are continually changing and becoming more demanding and interlocked.

The supplying utility cannot be expected to provide a perfect power supply because many of the causes of power supply disturbances are beyond the control of the utility. For example, automobiles hit poles, animals climb across insulators, lightning strikes overhead lines, and cyclones, hurricanes, and high winds blow tree branches and other debris into lines.

Lightning, wind, and rain produced by thunderstorms cause power failures in the form of power interruptions and transients. Fig 1 is useful in determining the probability of such power failures depending upon the user's geographic location. However, most utilities or commercial power companies, through the use of proper shielding and surge arrestors, can provide service that is of equal reliability in any area.

Fig 2 shows the density of tornadoes in several states. These violent local storms take their toll on power systems, as do more widely located storms such as hurricanes, ice and snow storms, and floods. The weighted consideration must be based on their frequency and severity in the user's location.

Although there is less chance for an interruption on an underground system, and although an underground system has fewer dips, the duration of any interruption which occurs may be much longer because of the longer time required to find where a cable failure occurred and to make repairs.

Even the operation of protective devices can cause power supply disturbances. As an example, overcurrent and short-circuit protective devices require excessive current to operate and will be accompanied by a voltage dip on any line supplying the excessive current until the device opens to clear the fault. Devices opening to clear a lightning flashover with subsequent reclosing cause a momentary interruption.

Utility companies have little practical control over, and thus cannot accept the responsibility for, disturbances on their system. This is illustrated by the text taken from a typical utility sales contract: "The power delivered hereunder shall be

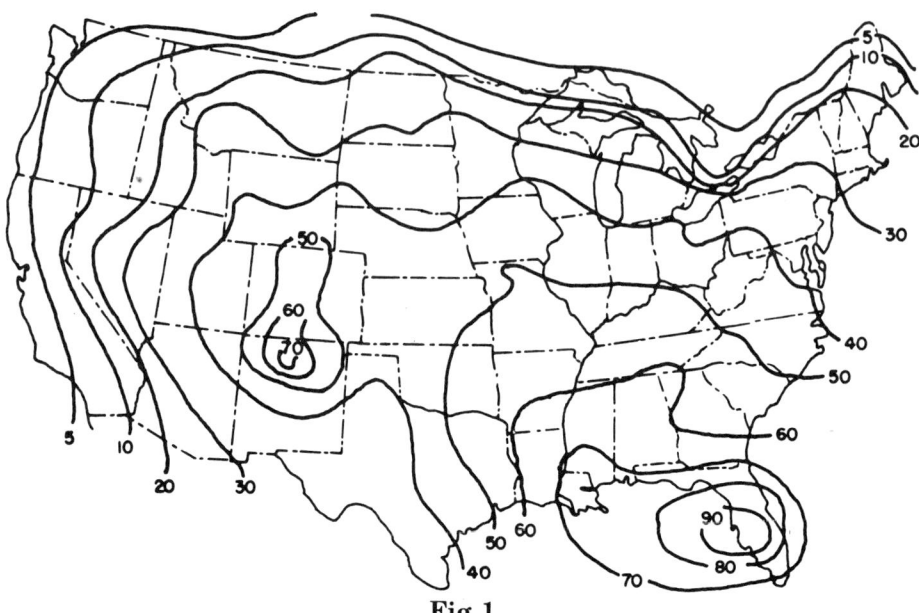

Fig 1
Average Number of Thunderstorm Days per Year

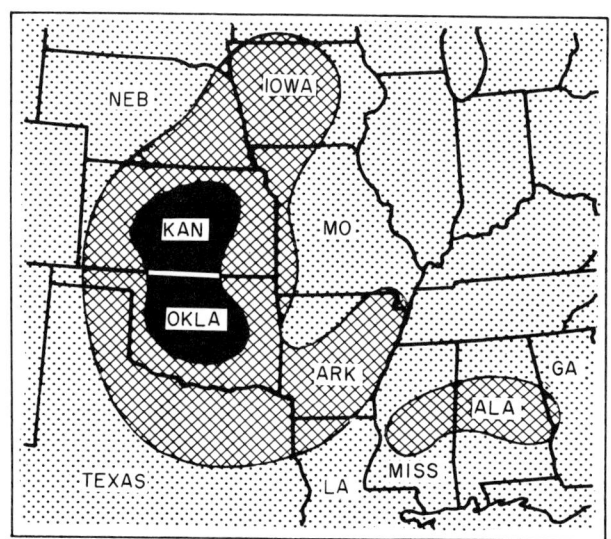

Fig 2
Approximate Density of Tornadoes
Cross-hatched areas—100-200 in 40 years;
black area—over 200 in 40 years
(See 3.15.1, [1].)

three-phase alternating current at a frequency of approximately 60 Hz and at a nominal voltage of 208Y/120 V. Except for temporary periods of abnormal operating conditions, variations from normal voltage shall not exceed 5 percent up or down. The utility will use reasonable diligence to provide a regular and uninterrupted supply of power, but in case such supply should be interrupted for any cause, the utility shall not be liable for damages resulting therefrom."

It is obvious that utility companies are unwilling to make power quality or continuity guarantees, and rightly so.

In several instances, utility companies have recognized the limitations of their power quality and have offered auxiliary equipment with special purchase agreements to satisfy the needs of those who use sensitive utilization equipment.

Utility companies are by no means the only source of power system disturbances. Disturbances and outages also occur in the plant system through the loss of power due to short circuits in wiring and failures of local generation including emergency and standby equipment. "Noise" is generated in otherwise acceptable electric power by motors, welders, switches, semiconductor-controlled rectifier (SCR) gating, dielectric heating, arcing short circuits, and a myriad of other sources.

Ordinary power meters and instrumentation cannot be used to measure power transients or disturbances. Recording devices with extremely fast response must be used to detect, measure, and record disturbance magnitudes and durations. Most disturbances will not only vary within a given 24 h period, but depending on the geographical area, could be subject to seasonal variations. Usually there are no simple methods available to record, diagnose, and classify the nature of the power problems.

Any meaningful data with regard to power quality, or lack of it, will be obvious only after conducting a thorough and detailed measurement recording and analysis program. Such a program would include the monitoring of an incoming power line over a representative period of time. Since much of the instrumentation is costly (typically $3000 to $16 000) in relation to its limited duration usage, rental arrangements are often a popular choice.

Since ideal power quality and continuity can seldom be obtained from the supplying utility, this text has been prepared to indicate how the effects of power supply disturbances may be reduced to acceptable levels or even eliminated. These reduction methods include the following:

(1) Modification of the design of utilization equipment so as to be impervious to power disturbances and discontinuities

(2) Modification of the prime power distribution system to be compatible with utilization equipment

(3) Modification of both systems and equipment to meet a criterion that is realistic for both

(4) Interposing a continuous electric supply system between the prime source and the utilization equipment. This will function as a "buffer" to external sources of transients, but depending on the design, could increase the magnitude of load induced disturbances.

Almost any significant deviation from normal power parameters may be capable of causing problems with some electronic equipment. Steady or slow deviations which exceed the product design range of line voltage or frequency can affect the shaft speed of motors, the force and speed of actuators, and the conversion of alternating voltage into regulated dc voltage for electronic circuit operation. As line voltage approaches product design limits, this often reduces the ability of the product to sustain, without incident, a line voltage transient as described in the following paragraphs. Of course some electronic circuits are more susceptible than others. This depends upon specific application and design.

Most frequent among excursions from normal line conditions are those classed as voltage transients. These often contain an initial fast voltage rise or fall time (sometimes oscillatory) followed by a slower, longer duration change. Thus one transient event may contain both fast impulses or noise and a slower change in voltage.

Fast voltage impulses and noise generally have little direct effect upon the smooth, even flow of power to such devices as motors and ac to dc power conversion equipment. The flywheel effect of motors and the energy storage in filters for reducing radio frequency interference and ripple in rectified dc output current of ac to dc converters keep the fast voltage changes from affecting their outputs directly. Excessively large impulses or noise bursts and inadequate filtering may result in component voltage overstress or in the premature triggering of control circuit elements such as SCRs and TRIACs. Otherwise, very short duration impulses and high frequency noise have little effect upon electronic circuits via power supply paths. More often these disturbances reach sensitive electronic circuits by more subtle paths through circuit grounds where power and signal ground circuit paths intersect. This dictates that care be taken in the location of connections and routing of ground conductors in addition to the more obvious precautions taken to reduce the number and amplitude of transients.

Without a good power disturbance monitor to measure and record disturbances continuously, there is a tendency to make judgments regarding disturbances in the power source in terms of their effects upon various electrical loads. When the lights go out and the electrical equipments all stop working, this is usually a fairly good indication of a power interruption. However, if the lights merely flicker or the electronic equipment malfunctions, it is difficult to judge whether there has been a severe change in voltage for a very short time or a very small voltage change for a much longer time. One cannot determine without a disturbance monitor whether disturbance was unusually severe or the electronic equipment was unusually susceptible to the disturbance. Without the necessary features in a disturbance monitor, one cannot tell whether the source of the disturbance was external to the load equipment or was the load equipment itself.

For practical reasons and convenience, power disturbances should be defined in terms which are related to practical methods of measurement. When power disturbance limits are given in the specifications for electrical and electronic equipment, use of the same terminology and definitions used in measurement will make it easier to resolve problems, and in some cases will help identify the responsibility.

For purposes of later discussions, disturbances in ac power may be classified as deviations in one or more of the following:
 (1) Voltage
 (2) Wave shape
 (3) Frequency
 (4) Phase relationships

Of these, the most frequently encountered deviations occur in voltage as follows:
 (1) Steady-state values (slow average), including unbalance
 (2) Outages and interruptions.

GENERAL NEED GUIDELINES

(3) Surges and sags
(4) Impulses and noise

NOTE: The term "transients" applies loosely to items (3) and (4).

The steady-state value is measured with a true rms-actuated, rms-reading voltmeter or equivalent. The instrument should be damped or the readings otherwise averaged over a 5 to 10 s averaging time so that a succession of readings will indicate the gradual, long-term changes in averaged rms voltage level. This is also known as slow-average rms voltage. Most deviations in steady-state voltage are caused by voltage drops in power lines, transformers, and feeders as load is increased. When step voltage corrections are made by such means as transformer tap changing or by adding or removing capacitors, transients will be generated at the time of the step change. In addition, the change which corrected an undervoltage condition as load increased will have to be reversed later when load decreases. Otherwise excessive overvoltage may result. Voltage is usually changed gradually in many small incremental steps. Occasional large step changes can occur.

The worst steady-state voltage deviations are likely to occur in areas where the total load approaches and even exceeds the capacity to generate or distribute power. In spite of short time overload ratings, it becomes necessary to reduce load during periods of peak loading such as during heat waves when air conditioning is added to all the other normal loads. Under such conditions, not only may there be an unusually large voltage drop between power source and load, but there may also be a planned voltage reduction to relieve utility system loading. This voltage reduction is known as a "brownout." With reductions of 3 percent, 5 percent, or even 8 percent at points where voltage is regular, the voltage reduction at load points may be an additional 5 to 10 percent.

The most common protection against brownouts is some form of voltage regulation, preferably one which has sufficiently low internal impedance and fast response time to avoid disturbances created by load changes and by phase controlled load regulation (where used).

When a planned brownout fails to relieve the utility power system overload enough to supply energy at reduced voltage to all users, one remaining alternative is to shed loads in a rotating sequence known as a "rolling blackout." Under some emergency plans, some noncritical loads would be shed for the duration of the overload. A few known critical loads would not be disturbed. The remainder would be subjected to power interruptions lasting 10 to 20 min each on a rotating basis among groups of subscribers, sufficient to relieve the overload by means of the rolling blackout.

Unequal loading on polyphase lines or single-phase three-wire lines is often the cause of voltage imbalance. Voltage control devices generally operate to regulate the average of the phase voltages, or sometimes just one of the phase voltages, on the assumption that the other phases will be equal to it. The most common corrective measure is to balance the loading among the phases. However, power sources with high internal impedance are more critical in a requirement to distribute loads evenly in order to avoid excessive voltage imbalance. Selection of a delta-connected primary with wye-connected secondary rather than four-wire wye input and output will help distribute unequal phase loading.

Large computer installations with an uninterruptable power supply (UPS) capable of supplying the computer's power needs for 5 to 15 min would need supplementary standby power to operate auxiliary building services, such as air conditioning, and to replenish the storage battery energy in order that the computer system could continue to remain functional through an extended power outage. Diesel or gas-turbine-driven generators fill this need.

Nonpermanent departures from the normal line voltages and frequency can be classified as disturbances. Disturbances include impulses, noise, transients, and even some changes in frequency or sudden phase shifts during synchronizing operations. Although the frequency and phase shift events may rarely be encountered when large loads are switched in power networks, these may frequently be encountered in small independent power sources during synchronizing and switching of loads from one source to another.

The more usual disturbance involves line voltage impulses, noise, transients, steady-state voltage change, or some combination of these. Most disturbances on a power system are of short duration. Studies indicate that 90 percent are less than 1 s in duration. Some 80 to 85 percent involve only one phase of a three-phase system. The following is an attempt to group disturbances in relation to their duration and possible causes:

1 s to 1 min. Usually attributed to severe faults accompanied by 50 to 100 percent voltage loss on one or more phases, these disturbances often result in an outage on some circuit. Faults often involve all three phases and may be the result of a downed pole, a tree or crane in the line, a breaker lockout, or an in-line fuse blowing. If the critical load is on the cleared side of the fault, the disturbance becomes an outage. If it is on the power source side of the fault clearing device, the normal voltage may be restored.

10–40 Cycles. These disturbances are surges and sags due to operation of relatively slow-speed breakers, reclosures in clearing faults on adjacent circuits, tap changing of in-line transformers and regulators, and the starting of motors, either across the line or reduce voltage start.

0–8 Cycles. The disturbances are surges, but more often sags, caused by a fault and subsequent fuse action or high-speed circuit breaker operation on adjacent circuits. Switching surges and inrush current transients caused when energizing electrical devices create short-duration voltage sags and surges. Single-phase loads on a multiline power source (single- or three-phase) create voltage surges on the unloaded phases if they cause sags on the loaded phase.

.001–1 Cycle. These disturbances are short-duration surges and sags caused by lightning arresters, load and capacitor switching, and short-duration faults. Any disturbances lasting less than 1 cycle may be difficult to compare directly with those lasting longer. Measurement of voltages for periods longer than one cycle is conveniently done in terms of rms values (the root mean square of all values during the cycle). Since most utilization voltages are supplied via a transformer in series with the power source, it is unlikely that a voltage surge at its output will exceed 130 to 150 percent of nominal volt-seconds (the area under a half sine wave). Core saturation in the transformer tends to limit this. Greater input volt-seconds which would push the magnetic core into deeper saturation would result in overcurrent tripping. The net result is that the transformer can pass very high voltages if they are of short duration, but the longer they persist, the lower the voltage must be.

Less than .001 Cycle. These disturbances are generally classed as impulses. (They originate as the fast portions at the beginning or end of switching transients which may have much longer durations. They may be associated with disturbances of all kinds, the more severe of which are natural lightning, electrostatic discharge, and the switching of near-by loads sharing the same power feeder.)

Impulses may be of either polarity, single fast-rise and fast-fall time event, or they can be damped-oscillatory in form. There can be bursts of multiple impulses. They can be synchronized with the power frequency so as to occur at the peak of each voltage wave (arcing fault, for example) or they can occur at random.

Impulses are measured in most instrumentation systems as voltage peak deviation from the sine wave voltage. This is usually accomplished by use of a high-pass filter (simple capacitor–resistor combination) to block the power frequency voltage. If a positive-going impulse occurs at the positive peak of a sine wave, the maximum instantaneous voltage will be the sum of the two. If not measured or expressed in volts, impulse amplitude limits are expressed in terms of the nominal voltage peak. For example, an impulse of "3 times nominal" referred to a 120 V rms nominal voltage would imply an impulse 3 times as great as the peak value of the sine wave. This is $\sqrt{2} \times 120 \times 3 = 509$ V peak value for the impulse. If this occurred at the peak of the sine wave and was in the same direction, the two would add to give $509 + (\sqrt{2} \times 120) = 679$ V peak.

A fast impulse applied to one conductor will tend to couple to the other nearby conductors as it travels. Its amplitude and rise time will also decrease as it travels. By the time that an impulse travels more than 20 or 30 ft, it is likely there will be considerable common mode signal in the conductor bundle unless the signal started as a balanced signal over a balanced line (equal positive and negative pulse signals on each of a balanced wire pair). In many sensitive electronic units, common mode impulses and high frequency noise from power source disturbances are a major cause of malfunctions. Isolating transformers placed close to the electronic equipment and the physical routing of grounding conductors to minimize intersection of power and low voltage level signal ground paths will often reduce or eliminate power-related transient noise problems.

Electrical noise can have much the same effect as impulses upon sensitive electronic equipment and requires similar measures to cope with it. In digital equipment, digital circuits generally operate in one of two (binary) states. Its circuits are relatively insensitive to external disturbances most of the time. However, at transition and clock times, a noise signal which otherwise could not influence a logic state might alter it and corrupt the signal. The statistical probability may be small but still create sufficient trouble to make a computer operation unacceptable to a user. Impulses are generally discrete events, and if there are not too many of them in a day, they may escape notice as a cause of difficulty. Noise, on the other hand, may be more continuous or occur in bursts. Depending upon occurrence and intensity, noise may be a more severe disturbance in terms of the number and frequency of malfunctions it causes.

Noise may be random in frequency content but is more likely to have strong frequency structure, some components of which may be at high radio frequencies. If power and grounding conductor lengths coincide with wavelengths (or multiples of quarter wavelengths) of noise frequencies, resonant conditions can occur which will greatly amplify the noise signals.

Grounding is an essential part of power

sources and their connections to loads. In addition to making ground conductors comply with local codes for safety to personnel (keeping exposed metal at ground potential) and providing low impedance return paths so that circuit protection will clear faults, grounding must also serve as a constant potential signal reference from one part of the grounding system to another over a broad frequency range.

For much electrical equipment, 70 to 90 percent voltage protection is adequate, that is, transfer to emergency power when the line voltage drops to 70 percent of normal and retransfer when it returns to 90 percent. When specialized loads are used, the manufacturer's recommendation should be followed. For motor loads, the protection can be increased to provide 80 to 90 percent protection. Too close a voltage spread will cause many false or unnecessary transfers; too wide a spread will cause equipment damage and malfunctions. The length of time of the voltage dip is of paramount importance.

Justification of the expenditures necessary to fulfill "user needs" for emergency and standby power systems falls into three broad classifications:

(1) Mandatory installations to meet laws and regulations of federal, state, county, and city

(2) Maintenance of safety and health of people involved during a power failure

(3) Increased profits due to fewer and shorter power failures

Power for the safety of personnel and to prevent pollution of the environment may become increasingly mandatory. A continual check of laws and regulations must be maintained to be current with the requirements. Public Law Number 91-596, Williams-Steiger Occupational Safety and Health Act of 1970 (OSHA), may materially alter the electric power reliability and availability requirements as the full interpretation and application of its provisions are published. See 3.14.

Table 1 is a guide to state codes and regulations for emergency power systems in the United States. All latest codes and regulations for the area in which the industrial or commercial facility is located must be consulted and followed.

Table 2 lists the needs in thirteen general categories with some breakdown under each to indicate major requirements. Ranges under the columns "Maximum Tolerance Duration of Power Failure" and "Recommended Minimum Auxiliary Supply Time" are assigned based upon experience. Written standards have been referenced where applicable.

In some cases under the column "Type of Auxiliary Power System" both emergency and standby have been indicated as required. An emergency supply of limited time capacity may be used at a low cost for immediate or uninterruptible power until a standby supply can be brought on line. An example would be the case where battery lighting units come on until a standby generator can be started and transferred to the critical loads.

Following the table, the "General Need" listings are presented in greater detail with recommendations as to the type of equipment or system which should be used.

Users of this publication may wish to skip the detailed presentation of each "General Need" and go directly to Section 4, "Systems and Hardware." If so, care should be taken that all individual needs have been recognized and listed so that suitable power systems can be selected to meet all requirements.

Readers using this recommended practice may find that various combinations of general needs will require an in-depth

Table 1
Codes for Emergency Power by States and Major Cities

State/City	Does State-City Have Legislation?	What Type?	Hospitals	Nursing Homes	Schools	Theaters (Public Gathering Places)	Office Buildings	Hotels	Apartment Buildings	Airports	Fire and Police Stations	Water Treatment Plants	Sewage Treatment Plants	All Public Buildings, State	All Public Buildings, Commercial	Applicable Governing Agency
Alabama	Yes	1	E	E												2
Mobile	Yes	7	E	E	C,D	E	E	E	C,D		C				C	3,7
Alaska	Yes	2														4
Arizona	Yes	3	E	E	C,D	C,D	C,D	C,D	E	E				C,D	C,D	3,5
Phoenix	Yes	1														3
Arkansas	Yes	1,2														1
Little Rock	Yes	2														1,7
California	Yes	2	E	E	E	E	E	C,D	E	E	E	E	E	E	E	5
Berkeley	Yes	1,3	A,C,D	E	C,D	C,D	C,D	C,D	C,D	C,D	C,D	A,C,D	C,D	C,D	C,D	1,5
Fresno	Yes	3														7
Glendale	Yes	3	A,C,D	C,D	C,D	C,D	C,D	C,D	C,D	C,D	A,C,D	C,D	C,D	C,D	C,D	1,5
Long Beach	Yes	3	A,C,D	A,C,D	A,C,D	A,C,D	C,D	C,D	C,D	C,D	C,D					3
Los Angeles	Yes	4														4
Pasadena	Yes	City	A,C,D	C,D	C,D	C,D	C,D	C,D	C,D	C,D	C,D	B,C,D	B,C,D			5
San Diego	Yes	1,3	E	E	C	A,C,D	A,C	C	C			C,D	C,D			7
San Francisco	Yes	3,4														7
San Jose	Yes	City														1,7
Santa Ana	Yes	4				D										
Colorado	Yes	3,4														
Connecticut	Yes	4	A,C,D	A,C,D	A,C,D	A,C,D	A,C,D	A,C,D	A,C,D	A,C,D	A,C,D	A,C,D	A,C,D	A,C,D	A,C,D	1,6
Hartford	Yes	2	A,C,D	A,C,D	A,C,D	A,C,D	A,C,D	A,C,D	A,C,D	A,C,D	A,C,D	A,C,D	A,C,D	A,C,D	A,C,D	1
New Haven	Yes	2														
Delaware	No	2						E	C,D							
Florida	Yes	7	E	E	E	E	E	E								
Jacksonville	Yes	City	E	E	E	E	E									5
St. Petersburg	Yes	4,7									A	A	A	A	A	7
Tampa	Yes	1														1

Note: An explanation of the numbers and letters used is given at the end of the table.

Table 1 (Continued)

	Does State-City Have Legislation?	What Type?	Hospitals	Nursing Homes	Schools	Theaters (Public Gathering Places)	Office Buildings	Hotels	Apartment Buildings	Airports	Fire and Police Stations	Water Treatment Plants	Sewage Treatment Plants	All Public Buildings, State	All Public Buildings, Commercial	Applicable Governing Agency
Georgia Columbus	Yes	1	E	E	C,D	E	E	E	E	E	A,B	E	E	E	E	1
Savannah	Yes	1,4	E	B,C,D	B,C,D	B,C,D	C,D	B,C,D	C,D	C,D	C,D	C,D	C,D	C,D	C,D	7
Hawaii Honolulu	Yes	1,3	A,C	C	C	C	C	C	C	C	C	C	C	C	C	1
Idaho	Yes	3,4	A,C,D	A,C,D	A,C,D	A,C,D	A,C,D	A,C,D	A,C,D	A,C,D	A,C,D	A,C,D	A,C,D	A,C,D	A,C,D	1,2
Illinois	Yes	2	E	E	E	A,C,D	A,C,D	C	C,D	A,C,D		B	B	E	E	5
Chicago	Yes	City	E	E	C,D	C,D		A,C,D	A,C,D							2
Rockford	Yes	1	A,C,D	A,C,D	A,C,D	A,C,D	C,D	A,C,D	C,D					A,C,D	A,C,D	7
Indiana	Yes	4														1,5
Evansville	Yes	1	A,C,D	A,C,D	C,D	C,D	C,D	A,C,D	C					C,D	C,D	1
Ft. Wayne	Yes	1	A,C,D	C,D	C,D	C,D	C,D	A,C,D	C,D					C,D	C,D	1
Gary	Yes	2	C,D	C,D	C,D	C,D	C,D	C,D	C,D	C,D	C,D	B	B,C,D	C,D	C,D	3
Indianapolis	Yes	4,5	E	A,C,D	C,D	C,D	E	C,D	C,D	E	A,C,D			E	A,C,D	7
South Bend	No															5
Iowa	Yes	1,2	E	E	C,D	E	E	E	E						E	3
Kansas	Yes	3,4	E	E	C,D	C,D	E	E	E	C,D	C,D	C,D	C,D	E	E	7
Kansas City	Yes	4	A	A	C,D	C,D	C,D	C,D	E		A			A	C,D	7
Wichita	Yes	1	E	E	E	C,D	C,D	C,D	C,D	C,D	C,D	C,D	C,D	C,D	C,D	1
Kentucky	No															124
Louisiana	Yes	1,4	A,C,D	A,C,D	C,D	C,D	C,D	C,D	C,D	C,D	C,D	C,D	C,D	C,D	C,D	7
Baton Rouge	Yes	4	E	E	C,D	C,D	C,D	C,D	C,D	C,D	A	C,D	C,D	A	C,D	1
New Orleans	Yes	1	A	A	C,D	C,D	C,D	A,C,D	C,D		C,D	C,D	C,D	C,D	C,D	2,4
Maine	Yes	2	C,D	C,D	C,D	A,C,D	C,D	A,C,D	A,C,D	A,C,D	A,C,D	A,C,D	A,C,D	A,C,D	A,C,D	3
Maryland Baltimore	Yes	City	A,C,D	A,C,D	A,C,D	A,C,D	C,D	A,C,D	A,C,D	A,C,D	A,C,D	A,C,D	A,C,D	A,C,D	A,C,D	1,5
Massachusetts Boston	Yes	1			A,D	C,D	C,D	D	C,D						C,D	1
Cambridge	Yes	2	A,D	C,D							A,C,D	A,C,D	A,C,D	A,C,D		4,7
New Bedford	Yes	2														7

Note: An explanation of the numbers and letters used is given at the end of the table.

Table 1 (Continued)

	Does State-City Have Legislation?	What Type?	Hospitals	Nursing Homes	Schools	Theaters (Public Gathering Places)	Office Buildings	Hotels	Apartment Buildings	Airports	Fire and Police Stations	Water Treatment Plants	Sewage Treatment Plants	All Public Buildings, State	All Public Buildings, Commercial	Applicable Governing Agency
Springfield	Yes	2	E	A,C,D	A,C,D	A,C,D	A,C,D	A,C,D	A,C,D		A,C,D			A,C,D	A,C,D	5
Worcester	No	—	E													1,6
Michigan	Yes	1,2	E	E	A,C,D	A,C,D	A,C,D	A,C,D	E		C,D	B	B	E	E	5
Detroit	Yes	1,5	E	E	A,C,D	A,C,D	A,C,D	E	E	E	E				D	7
Flint	Yes	1,4,5	E	E	A,C,D	A,C,D	A,C,D									7
Lansing	Yes	2	E	E	A,C,D	A,C,D	E									3
Minnesota	Yes	3			E	E	E	E	E	A,C,D	E	A,C,D	A,C,D	E	E	7
St. Paul	Yes	1,3,4	E	A,C,D	E	C,D	C,D	C,D	C,D	C,D	C,D	C,D	C,D	C,D	C,D	4
Mississippi	Yes	1															1
Jackson	Yes	5															
Missouri	Yes	—															4
Kansas City	Yes	3,4			A,C,D	A,C,D	C,D	A,C,D	C,D	A,C,D	E	C,D	C,D	C,D	C,D	1
Montana	Yes	1		E	E	C,D	C,D	C,D	C,D	C,D	C,D			C,D	C,D	1
Nebraska	Yes	1		E	E	C,D	C,D	C,D	C,D	C,D	C,D			C,D	C,D	7
Lincoln	Yes	3,4		E	C,D	E	E	E	E	E		A,B	A,B	E	E	1
Nevada	Yes	1	E	E	E	E	C,D	C,D	C,D							6
New Hampshire	Yes	1	E	E	C,D	C,D	C,D	C,D	C,D	A,C,D		A,B	A,B	C,D	C,D	2,3
New Jersey	Yes	1,2	E	E	E	E	E	E	E	A,C,D	E	E	E	E	E	3
New York	Yes	2	E	E	A,C,D	A,C,D	A,C,D	A,C,D	A,C,D	A,C,D	A,C,D			A,C,D	A,C,D	6
Albany	Yes	2	E	E	A,C,D	A,C,D	C	A,C,D	A,C,D	A,C,D	E	C,D	C,D			7
New Mexico	Yes	1,4	A,C,D	A,C,D					D	D		C	C			4,5
North Carolina	Yes	2	C,D	D	D	D	D	D	D		E			E		5
North Dakota	Yes	1	E	E	E	E	E	E	E							7
Ohio	Yes	6															7
Akron	Yes	1,6			C,D						E			E	C	7
Canton	Yes	4,6														5
Cincinnati	Yes	6														1,4
Cleveland	Yes	City														
Columbus	Yes	6														
Youngstown	Yes	6														7
Oklahoma	No	—															

Note: An explanation of the numbers and letters used is given at the end of the table.

Table 1 (Continued)

		What Type?	Hospitals	Nursing Homes	Schools	Theaters (Public Gathering Places)	Office Buildings	Hotels	Apartment Buildings	Airports	Fire and Police Stations	Water Treatment Plants	Sewage Treatment Plants	All Public Buildings, State	All Public Buildings, Commercial	Applicable Governing Agency
Oregon	Yes	3,4	A,C,D	A,C,D	A,C,D	A,C,D	A,C,D	A,C,D	A,C,D	A,C,D		A,C,D	A,C,D	A,C,D	A,C,D	3
Portland	Yes	1	A,C,D	C,D	C,D	C,D	C,D	C,D	C,D	C,D	C,D	A,C,D	A,C,D	C,D	C,D	1
Pennsylvania	Yes	1	E	E	E	E	C,D	C,D	C,D	C,D	C,D	E	E	C,D	C,D	6
Rhode Island	Yes	2	E	E	E	E	A,C,D	A,C,D	A,B,C	C,D		E	E	E	C,D	1
South Carolina	Yes	1,4,7	E	E	E	E	A,C,D	A,C,D	A,C,D			E	E	E	C,D	5
South Dakota	Yes	1,4	C,D	C,D	C,D	C,D	C,D	C,D	C,D	C,D	C,D	C,D	C,D	C,D	C,D	2,5
Tennessee	Yes	2														1
Texas	Yes	2														2
Amarillo	Yes	3	C,D	C,D	C,D	C,D	C,D	C,D	C,D	C,D	C,D	C,D	C,D	C,D	C,D	3
Corpus Christi	Yes	4	E	E		E	C,D	C,D	C,D	C,D				C,D	C,D	1,7
Dallas	Yes	3,4	C	C	C	C	A,B,C	A,B,C	A,B,C	C	C	C	C	C	C	7
Fort Worth	Yes	3,4	E	C,D	C,D	C,D	C,D	C,D	C,D	C,D	E	C	C	E	C	7
San Antonio	Yes	4	E	E	A,C,D	A,C,D	A,C,D	A,C,D	A,C,D	A,C,D	A,C,D	A,C,D	A,C,D	A,C,D	A,C,D	3
Wichita Falls	Yes	7	C,D	C,D	C,D	C,D	C,D	C,D	C,D	C,D	C,D	C,D	C,D	C,D	C,D	7
Utah	Yes	4														
Salt Lake City	No	4	E	E	A,C,D	A,C,D	C,D	C,D	C,D	A,C,D	C,D	C,D	C,D	C,D	C,D	1,7
Vermont	Yes	1,2	E	E			C,D	C,D	C,D							1
Virginia	Yes	1,2	E	E		E	C,D	E	E							6
Washington	Yes	1,3	E	E	E	C,D	C,D	C,D	C,D		C,D	A,B	A,B	A,B	C,D	1,6
Seattle	Yes	2														1,3
West Virginia	Yes	2	E	E	E	C,D	C,D	C,D	C,D			C,D	C,D	C,D	C,D	6
Wisconsin	Yes	1	E	E	E	C,D	C,D	C,D	C,D			C,D	C,D	C,D	C,D	1
Wyoming																

Legislation Code
1. National Fire Protection Association
2. State
3. Uniform Building Code
4. National Electrical Code
5. Building Officials and Code Administration
6. Ohio Building Code
7. Southern State Building Code

Governing Agency
1. Fire Marshall
2. Department of Public Health
3. Local Government Units
4. Public Safety
5. State Building Standard Commission
6. Department of Labor
7. Inspection Department

Power Source
A. Emergency Power
B. Standby Power
C. Exit Lighting
D. Egress Lighting
E. All

Note: Table 1 courtesy of Electrical Generating Systems Marketing Association (April 1975).

Table 2
Condensed General Criteria for Preliminary Consideration

Section	General Need	Specific Need	Maximum Tolerance Duration of Power Failure	Recommended Minimum Auxiliary Supply Time	Type of Auxiliary Power System Emergency	Type of Auxiliary Power System Standby	System Justification
3.1	Lighting	Evacuation of personnel	Up to 10 s, preferably not more than 3 s	2 h	×		Prevention of panic, injury, loss of life Compliance with building codes and local, state, and federal laws Lower insurance rates Prevention of property damage Lessening of losses due to legal suits
		Perimeter and security	10 s	10–12 h during all dark hours	×	×	Lower losses from theft and property damage Lower insurance rates Prevention of injury
		Warning	From 10 s up to 2 or 3 min	To return to prime power source	×		Prevention or reduction of property loss Compliance with building codes and local, state, and federal laws Prevention of injury and loss of life
		Restoration of normal power system	1 s to indefinite depending on available light	Until repairs completed and power restored	×	×	Risk of extended power and light outage due to a longer repair time
		General lighting	Indefinite; depends on analysis and evaluation	Indefinite; depends on analysis and evaluation		×	Prevention of loss of sales Reduction of production losses Lower risk of theft Lower insurance rates

Table 2 (Continued)

Section	General Need	Specific Need	Maximum Tolerance Duration of Power Failure	Recommended Minimum Auxiliary Supply Time	Type of Auxiliary Power System — Emergency	Type of Auxiliary Power System — Standby	System Justification
		Hospitals and medical areas	Uninterruptible to 10 s NFPA No 76A–1977 and 101–1976 allow 10 s for alternate power source to start and transfer	To return of prime power	×	×	Facilitate continuous patient care by surgeons, medical doctors, nurses, and aids Compliance with all codes, standards, and laws Prevention of injury or loss of life Lessening of losses due to legal suits
		Orderly shutdown time	0.1 s to 1 h	10 min to several hours	×		Prevention of injury or loss of life Prevention of property loss by a more orderly and rapid shutdown of critical systems Lower risk of theft Lower insurance rates
3.2	Startup power	Boilers	3 s	To return of prime power	×		Return to production Prevention of property damage due to freezing Provision of required electric power
		Air compressors	1 min	To return of prime power		×	Return to production Provision for instrument control
3.3	Transportation	Elevators	15 s to 1 min	1 h to return of prime power		×	Personnel safety Building evacuation Continuation of normal activity
		Material handling	15 s to 1 min	1 h to return of prime power		×	Completion of production run Orderly shutdown Continuation of normal activity
		Escalators	15 s to no requirement for power	Zero to return of prime power		×	Orderly evacuation Continuation of normal activity
		Conveyors	15 s to 1 min	As analyzed and economically justified		×	Completion of production run Completion of customer order Orderly shutdown Continuation of normal activity

GENERAL NEED GUIDELINES

IEEE
Std 446-1980

Table 2 (Continued)

Section	General Need	Specific Need	Maximum Tolerance Duration of Power Failure	Recommended Minimum Auxiliary Supply Time	Type of Auxiliary Power System Emergency	Type of Auxiliary Power System Standby	System Justification
3.4	Mechanical utility systems	Water (cooling and general use)	15 s	½ h to return of prime power		×	Continuation of production. Prevention of damage to equipment. Supply of fire protection
		Water (drinking and sanitary)	1 min to no requirement	Indefinite until evaluated		×	Providing of customer service. Maintaining personnel performance
		Boiler power	0.1 s	1 h to return of prime power	×		Prevention of loss of electric generation and steam. Maintaining production. Prevention of damage to equipment
		Pumps for water, sanitation, and production fluids	10 s to no requirement	Indefinite until evaluated		×	Prevention of flooding. Maintaining cooling facilities. Providing sanitary needs. Continuation of production. Maintaining boiler operation
		Fans and blowers for ventilation and heating	0.1 s to return of normal power	Indefinite until evaluated	×		Maintaining boiler operation. Providing for gas-fired unit venting and purging. Maintaining cooling and heating functions for buildings and production
3.5	Heating	Food preparation	5 min	To return of prime power		×	Prevention of loss of sales and profit. Prevention of spoilage of in-process preparation
		Process	5 min	Indefinite until evaluated; normally for time for orderly shutdown, or to return of prime power		×	Prevention of in-process product damage. Prevention of property damage. Continued production. Prevention of payment to workers during no production. Lower insurance rates

35

Table 2 (Continued)

Section	General Need	Specific Need	Maximum Tolerance Duration of Power Failure	Recommended Minimum Auxiliary Supply Time	Type of Auxiliary Power System Emergency	Type of Auxiliary Power System Standby	System Justification
3.6	Refrigeration	Special equipment or devices which have critical warmup (cryogenics)	5 min	To return of prime power		×	Prevention of equipment or product damage
		Depositories of critical nature, (blood banks, etc)	5 min (10 s per NFPA No 76A–1977)	To return of prime power		×	Prevention of loss of material stored
		Depositories of noncritical nature, (meat, produce, etc)	2 h	Indefinite until evaluated		×	Prevention of loss of material stored Lower insurance rates
3.7	Production	Critical process power (sugar factory, steel mills, chemical processes, glass products, etc)	1 min	To return of prime power or until orderly shutdown		×	Prevention of product and equipment damage Continued normal production Reduction of payment to workers on guaranteed wages during nonproductive period Lower insurance rates Prevention of prolonged shutdown due to nonorderly shutdown
		Process control power	Uninterruptible (UPS) to 1 min	To return of prime power	×		Prevention of loss of machine and process computer control program Maintaining production Prevention of safety hazards from developing Prevention of out-of-tolerance products
3.8	Space conditioning	Temperature (critical application)	10 s	1 min to return of prime power	×		Prevention of personnel hazards Prevention of product or property damage Lower insurance rates Continuation of normal activities Prevention of loss of computer function

Table 2 (Continued)

Section	General Need	Specific Need	Maximum Tolerance Duration of Power Failure	Recommended Minimum Auxiliary Supply Time	Type of Auxiliary Power System Emergency	Type of Auxiliary Power System Standby	System Justification
		Pressure (critical) pos/neg atmosphere	1 min	1 min to return of prime power	×		Prevention of personnel hazards Continuation of normal activities Prevention of product or property damage Lower insurance rates Compliance with local, state, and federal codes, standards, and laws
		Humidity (critical)	1 min	To return of prime power		×	Prevention of loss of computer functions Maintenance of normal operations and tests Prevention of explosions or other hazards
		Static charge	10 s or less	To return of prime power	×		Prevention of static electric charge and associated hazards Continuation of normal production (printing press operation, painting spray operations)
		Building heating and cooling	30 min	To return of prime power		×	Prevention of loss due to freezing Maintenance of personnel efficiency Continuation of normal activities
		Ventilation (toxic fumes)	15 s	To return of prime power or orderly shutdown	×		Reduction of health hazards Compliance with local, state, and federal codes, standards, and laws Reduction of pollution
		Ventilation (explosive atmosphere)	10 s	To return of prime power or orderly shutdown	×		Reduction of explosion hazard Prevention of property damage Lower insurance rates Compliance with local, state, and federal codes, standards, and laws Lower hazard of fire Reduce hazards to personnel

Table 2 (Continued)

Section	General Need	Specified Need	Maximum Tolerance Duration of Power Failure	Recommended Minimum Auxiliary Supply Time	Type of Auxiliary Power System Emergency	Type of Auxiliary Power System Standby	System Justification
		Ventilation (building general)	1 min	To return of prime power		×	Maintaining of personnel efficiency; Providing make-up air in building
		Ventilation (special equipment)	15 s	To return of prime power or orderly shutdown	×		Purging operation to provide safe shutdown or startup; Lowering of hazards to personnel and property; Meeting requirements of insurance company; Compliance with local, state, and federal codes, standards, and laws; Continuation of normal operation
		Ventilation (all categories noncritical)	1 min	Optional		×	Maintaining comfort; Preventing loss of tests
		Air Pollution control	1 min	Indefinite until evaluated; compliance or shutdowns are options	×		Continuation of normal operation; Compliance with local, state, and federal codes, standards, and laws
3.9	Fire protection	Annunciator alarms	1 s	To return of prime power	×		Compliance with local, state, and federal codes, standards, and laws; Lower insurance rates; Minimizing life and property damage
		Fire pumps	10 s	To return of prime power		×	Compliance with local, state, and federal codes, standards, and laws; Lower insurance rates; Minimizing life and property damage
		Auxiliary lighting	10 s	5 min to return of prime power		×	Servicing of fire pump engine should it fail to start; Providing visual guidance for fire-fighting personnel

Table 2 (Continued)

Section	General Need	Specified Need	Maximum Tolerance Duration of Power Failure	Recommended Minimum Auxiliary Supply Time	Type of Auxiliary Power System Emergency	Type of Auxiliary Power System Standby	System Justification
3.10	Data processing	CPU memory tape/disc storage, peripherals	½ cycle	To return of prime power or orderly shutdown	×		Prevention of program loss Maintaining normal operations for payroll, process control, machine control, warehousing, reservations, etc
		Humidity and temperature control	5–15 min (1 min for water-cooled equipment)	To return of prime power or orderly shutdown		×	Maintenance of conditions to prevent malfunctions in data processing system Prevention of damage to equipment Continuation of normal activity
3.11	Life support and life safety systems (medical field, hospitals, clinics, etc)	X-ray	Milliseconds to several hours	From no requirement to return of prime power, as evaluated	×		Maintenance of exposure quality Availability for emergencies
		Light	Milliseconds to several hours	To return of prime power	×		Compliance with local, state, and federal codes, standards, and laws Preventing interruption to operation and operating needs
		Critical to life, machines, and services	½ cycle to 10 s	To return of prime power	×		Maintenance of life Prevention of interruption of treatment or surgery Continuation of normal activity Compliance with local, state, and federal codes, standards, and laws
		Refrigeration	5 min	To return of prime power		×	Maintaining blood, plasma, and related stored material at recommended temperature and in prime condition

Table 2 (Continued)

Section	General Need	Specified Need	Maximum Tolerance Duration of Power Failure	Recommended Minimum Auxiliary Supply Time	Type of Auxiliary Power System		System Justification
					Emergency	Standby	
3.12	Communication systems	Teletypewriter	5 min	To return of prime power		×	Maintenance of customer services Maintenance of production control and warehousing Continuation of normal communication to prevent economic loss
		Inner building telephone	10 s	To return of prime power	×		Continuation of normal activity and control
		Television (closed circuit and commercial)	10 s	To return of prime power		×	Continuation of sales Meeting of contracts Maintenance of security Continuation of production
		Radio systems	10 s	To return of prime power	×		Maintenance of security and fire alarms Providing evacuation instructions Continuation of service to customers Prevention of economic loss Directing vehicles normally
		Intercommunication systems	10 s	To return of prime power		×	Providing evacuation instructions Directing activities during emergency Providing for continuation of normal activities Maintaining security
		Paging systems	10 s	½ h	×		Locating of responsible persons concerned with power outage Providing evacuation instructions Prevention of panic

Table 2 (Continued)

Section	General Need	Specified Need	Maximum Tolerance Duration of Power Failure	Recommended Minimum Auxiliary Supply Time	Type of Auxiliary Power System Emergency	Type of Auxiliary Power System Standby	System Justification
3.13	Signal circuits	Alarms and annunciation	1 to 10 s	To return of prime power	×		Prevention of loss from theft, arson, or riot Maintaining security systems Compliance with codes, standards, and laws Lower insurance rates Alarm for critical out-of-tolerance temperature, pressure, water level, and other hazardous or dangerous conditions Prevention of economic loss
		Land-based aircraft, railroad, and ship warning systems	1 s to 1 min	To return of prime power	×		Compliance with local, state, and federal codes, standards, and laws Prevention of personnel injury Prevention of property and economic loss

system and cost analysis which will modify the recommended equipment and systems to best meet the specific requirements.

Small commercial establishments and manufacturing plants will usually find their requirements under two or three of the general need guidelines given in Section 3. Large manufacturers and commercial facilities will find that portions or all of the need guidelines given in Section 3 apply to their operations and justify or require emergency and backup standby electric power.

3.1 Lighting

3.1.1 *Introduction.* Evaluation of the quality, quantity, type, and duration of emergency or standby power for lighting is necessary for each particular application. The different types of systems have various degrees of reliability which should be considered in the selection of the proper system.

3.1.2 *Lighting for Evacuation Purposes.* Interruption of power to a normal lighting system may cause injury or loss of life. Emergency lighting for evacuation purposes must energize automatically upon loss of normal lighting. Where legally required [NFPA No 101, Life Safety Code (1973)] emergency systems are installed, lighting must be maintained for at least 1½ h if battery-powered unit equipment is used. Emergency lighting must provide enough illumination to allow easy and safe egress from the area involved. All exit lights, signs, and stairwell lights should be included in both the emergency lighting system and the normal lighting system. Design of the emergency lighting system should include consideration of lighting for seeing, by silhouetting, protruding machines or objects in aisles.

3.1.3 *Perimeter and Security Lighting.* Emergency or standby power for perimeter and security lighting may be deemed necessary to reduce risk of injury, theft, or property damage. The power for perimeter lighting may not be required until several minutes after failure of normal power. In order to maintain perimeter lighting throughout the dark hours, a system should be capable of supplying power for 10 to 12 h for every 24 h the normal power source is off. For this reason the unit battery equipment is not recommended for auxiliary perimeter lighting.

3.1.4 *Warning Lights.* Emergency power should be available for all warning lights such as aircraft warning lights on high structures, ship warning lights on edges of waterways, and other warning lights which act to prevent injury or property damage. The power source selected should be capable of supplying emergency power throughout the duration of the longest anticipated power failure; therefore, the unit battery type is normally not suited for this application.

3.1.5 *Health-Care Facilities.* Emergency lighting is of paramount importance in hospitals and similar institutions. The requirements for these areas are included in 3.11.

3.1.6 *Standby Lighting for Equipment Repair.* Standby power for lighting should be installed in areas where the most probable internal power system failures may occur and in the main switchgear rooms. This requirement is justified by the necessity of having enough light to repair the equipment which failed and caused the loss of normal lighting.

3.1.7 *Lighting for Production.* Interruption of power to a normal lighting system may cause serious curtailment or complete loss of production. Where there are no safety hazards or property damage associated with this need, the decision should be based on the economic evaluation of each particular application. Systems which provide power for emergency lighting may also provide high-level lighting to allow production to continue.

3.1.8 *Lighting to Reduce Hazards to Machine Operators.* A machine operator may be subjected to a high injury risk for the first few seconds after lighting has failed. Many machines present a safety hazard if suddenly plunged into darkness. Instantaneous emergency lighting is required for protection against this type of injury.

3.1.9 *Supplemental Lighting for High-Voltage Discharge Systems.* If mercury or other types of high-voltage discharge lighting are used for the regular system, consideration should be given to adding auxiliary lamps such as incandescent or fluorescent. Some high-voltage discharge lamps require a cooling period before they restrike the arc and a warm-up period before they attain full brilliance. The total time required for full illumination after a momentary power interruption ranges from 1 min for high pressure sodium to 20 min for metal halide and mercury vapor.

3.1.10 *Codes, Rules, and Regulations.* Many states and municipalities have adopted their own specific codes regarding emergency lighting, in addition to those set down by the following:

(1) Occupational Safety and Health Act of 1970 (OSHA); OSHA is charged with enforcing compliance and makes reference to NEC and NFPA

(2) The National Electric Code, NFPA No 70-1978, Article 700, sets forth the standard of practice for emergency lighting equipment with regard to installation, operation, and maintenance

(3) The Life Safety Code, NFPA No 101-1976 (ANSI), concerns itself with the specification of locations where emergency lighting is considered essential to life safety and specifics on exit marking

(4) Underwriter's Laboratories tests and approves equipment to uniform performance standards as established by UL 924.

3.1.11 *Recommended Systems.* For short time durations, primarily lighting for personnel safety and evacuation purposes, battery units are satisfactory. Where longer service and heavier loads are required, an engine- or turbine-driven generator is usually used, which starts automatically upon failure of the prime power source with the load applied by an automatic transfer switch [2].[1] It is generally considered that an average level of 0.4 footcandles is adequate where passage is required and no precise operations are expected [3].

Table 3 summarizes the user's needs for emergency and standby electric power for lighting by application and areas.

3.2 Startup Power

3.2.1 *Introduction.* Assume a "cold" boiler and a "dead" plant without electrical power or steam. From this premise several very important questions must be answered, such as the following:

(1) How will the plant be protected from freezing in cold weather? Even with gas heaters, will there be sufficient heat without fans and without interlocked make-up air units running?

(2) A steam turbine generator is on hand but without forced draft, induced draft, boiler feed water, flame detectors, or control power. How can it be started?

(3) A gas turbine generator has been installed, but how can this be started without bringing it up in speed with a small steam turbine, an electric motor, or

---

[1]Numbers in brackets correspond to those in the References at the end of this Section.

Table 3
Typical Emergency and Standby Lighting Recommendations

Standby*	Immediate, Short Term†	Immediate, Long Term‡

Security lighting
  Outdoor perimeters
  Closed circuit TV
  Night lights
  Guard stations
  Entrance gates
Production lighting
  Machine areas
  Raw materials storage
  Packaging
  Inspection
  Warehousing
  Offices
Commercial lighting
  Displays
  Product shelves
  Sales counters
  Offices
Miscellaneous
  Switchgear rooms
  Landscape lighting
  Boiler rooms
  Computer rooms
	Evacuation lighting
	  Exit signs
	  Exit lights
	  Stairwells
	  Open areas
	  Tunnels
	  Halls
	Miscellaneous
	  Standby generator areas
	  Hazardous machines
		Hazardous areas
		  Laboratories
		  Warning lights
		  Storage areas
		  Process areas
		Warning lights
		  Beacons
		  Hazardous areas
		  Traffic signals
		Health-care facilities
		  Operating rooms
		  Delivery rooms
		  Intensive care areas
		  Emergency treatment areas
		Miscellaneous
		  Switchgear rooms
		  Elevators
		  Boiler rooms

* An example of a standby lighting system is an engine-driven generator.
† An example of an immediate short-term lighting system is the common unit battery equipment.
‡ An example of an immediate long-term lighting system is a central battery bank rated to handle the required lighting load only until a standby engine-driven generator is placed on line.

other prime mover? Gas compressors may be necessary and also require prime movers of some type.

(4) Steam and electrically driven fire pumps are out of service. There may be no major fire protection until electric power or steam is restored.

(5) An uninterruptible power supply of sufficient capacity is probably not on hand: otherwise, steam and electric power would not be down.

These statements illustrate the fact that adequate startup power is one of the most important considerations in the original design of any plant. Millions of dollars worth of equipment could be standing idle in a time of critical need if no allowance had been made for starting the machines under unexpected conditions such as a major power outage.

3.2.2 *Example of System Utilizing Startup Power.* Starting major plant equipment without outside power is commonly referred to as a "black start" and is accomplished by using only the facilities available within the plant. One example of a system designed with "black start" capability, with a minimum electrical startup system, would be a large gas turbine driving a centrifugal compressor in the natural gas pipeline industry, where the high-pressure gas from the pipeline is used to drive expansion turbines and gas motors for cranking the turbine, operating pumps, and positioning valves. By utilizing the high-pressure gas for the large horsepower requirements, a small engine-driven generator, fueled by natural gas, is used to provide electric power for turbine accessories, battery charging, lighting, and powering other critical loads. When the turbine is running, a shaft-driven generator provides larger quantities of electric power for all station requirements; the small generator is placed on "standby."

3.2.3 *Lighting.* In the design of the startup power system, first consideration should be given to installing battery-operated lights in the vicinity of the standby power source and switch-gear.

3.2.4 *Engine-Driven Generators.* The battery-cranking power for the engine may also be used for some lights. Cranking may also be accomplished by compressed air supplied to a tank by the plant compressed-air system and prevented from leaving the tank into the plant system by a check valve.

The engines may be sized for short-term operation if some type of continuous power generation can be brought onto the line after startup. If no generation as a prime source of electric power has been installed, the diesel or engine generator should be sized for supporting all electrical needs for the generator auxiliaries, boilers, critical emergency lights, fire signals, exit lights, and other items listed in Table 2.

3.2.5 *Battery Systems.* Special consideration should be given to the design of the plant battery system or uninterruptible power supply, allowing (1) adequate battery capacity to provide power for the necessary startup control systems, following a programmed safe stop; and (2) special disconnecting devices to automatically disconnect large power-consuming systems from the battery system, when possible, to conserve battery capacity for restart.

With capacity for these minimum facilities in operation, consideration may then be given to installing sufficient capacity at the same time to support additional justified needs.

3.2.6 *Other Systems.* Mobile equipment may suffice if it can reasonably be assumed to be available when needed. (Who has the highest priority when all have the need?) An alternate standby public utility line may also be available at a low cost from a separate source of supply. A neighboring plant with live generation may assist in an emergency.

3.2.7 *System Justification.* A definite workable plan with the proper equipment should be evaluated, and action taken as justified, prior to the need.

3.3 Transportation

3.3.1 *Introduction.* This topic covers the moving of people and products by methods which depend upon electric power and the importance of maintaining power ranges from desirable to critical.

3.3.2 *Elevators.* Where two or more elevators are in use in buildings three or more stories high, the elevators or banks of elevators should be connected to separate sources of power. There are situations where standby power is required for all elevators within 15 s. Savings may be made by supplying power during outages of the normal supply to one half the elevators installed, providing the traffic can be rerouted and the capacity of the elevators is adequate. Power must be transferred to the second banks of elevators within 1 min or so of the prime power loss to clear stalled elevators. Power may be left on this bank until normal power returns.

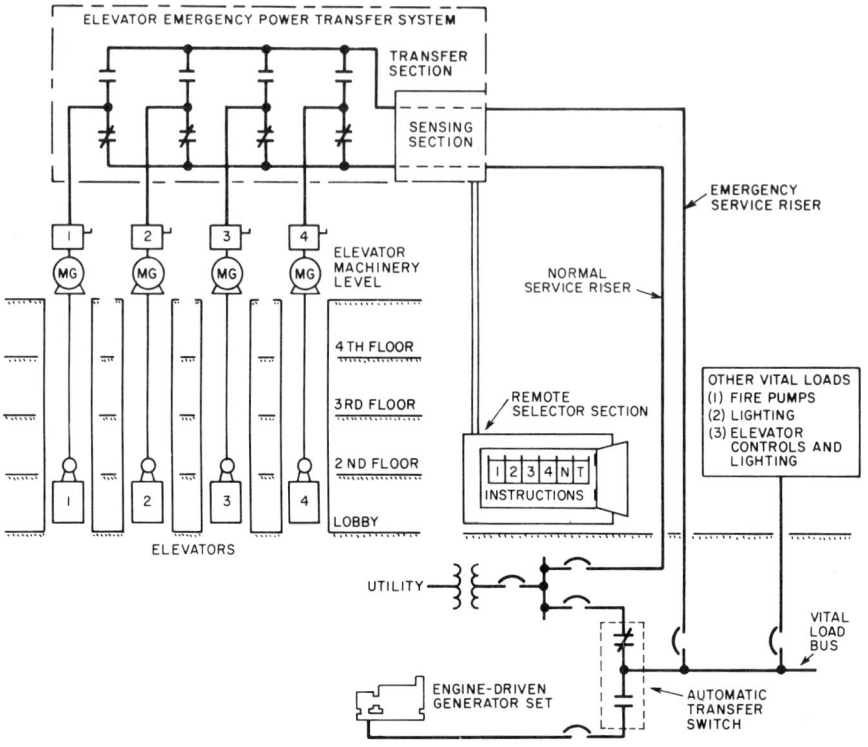

Fig 3
Elevator Emergency Power Transfer System

Where elevator service is critical for personnel and patients, it is desirable to have automatic power transfer with manual supervision. Operators and maintenance men may not be available in time if the power failure occurs on a weekend or at night.

(1) *Typical Elevator System.* Fig 3 shows an elevator emergency power transfer system whereby one preferred elevator is fed from a vital load bus through an emergency riser while the rest of the elevators are fed from the normal service. By providing an automatic transfer switch for each elevator and a remote selector station, it is possible to select individual elevators, thus permitting complete evacuation in the event of power failure. The engine generator set and emergency riser need only be sized for one elevator, thus minimizing the installation cost. The controls for the remote selector, automatic transfer switches, and engine starting are independent of the elevator controls, thereby simplifying installation.

(2) *Regenerated Power.* Regenerated power is a concern for motor generator type elevator applications. In some elevator applications, the motor is used as a brake when the elevator is descending and generates electricity. Electric power is then pumped back into the power

source. If the source is commercial utility power, it can easily be absorbed. If the power source is an engine-driven generator, the regenerated power can cause the generating set and the elevator to overspeed. To prevent overspeeding of the elevator, the maximum amount of power that can be pumped back into the generating set must be known. The permissible amount of absorption is approximately 20 percent of the generating set's rating in kilowatts. If the amount pumped back is greater than 20 percent, other loads must be connected to the generating set, such as emergency lights or "dummy" load resistances. Emergency lighting should be permanently connected to the generating set for maximum safety. A dummy load can also be automatically switched on the line whenever the elevator is operating from an engine-driven generator.

3.3.3 *Conveyors and Escalators.* Escalators and personnel conveyors may require emergency power since physically handicapped persons ride up and would have great difficulty walking down, even though a normal person would be able to do so were the power off.

Those who have used conveyors or elevators in a small way for a time and who have continued to grow, should be wary and check their needs for standby power. To feed a few cattle or gather a few eggs using conveyors may have once eased someone's labor, but labor could always fill in if needed. As operations become larger and larger, there is a point where an outage of electric power could produce a disaster. The gathering of 250 000 eggs a day or the feeding of 10 000 head of cattle is dependent upon power when needed.

3.3.4 *Other Transportation Systems.* Power for charging equipment for battery-powered vehicles is usually a noncritical requirement. Time available for the emergency generation to come on line varies from several minutes to several hours.

A few examples of transportation systems which may need standby power are listed below:

(1) Conveyors for raw materials through to finished goods
(2) Warehouse high-stacking loading and unloading equipment and conveyors delivering finished goods to shipping facilities
(3) Slurry pumps for long pipe lines
(4) Livestock feeder conveyors

3.4 Mechanical Utility Systems

3.4.1 *Introduction.* Often the need for mechanical utilities is as great as that for electric power. These are interdependent. We can speak of the utility systems as a whole since, to most managers and corporate officials, these are all united in a group under the heading "Utilities."

3.4.2 *Typical Utility Systems for which Reliable Power May Be Necessary.* Mechanical utility systems comprise the following services for which reliable electric power is usually required:

(1) Compressed air for pneumatic power
(2) Cooling water (including return pumps, pressure pumps, tower fans)
(3) Well water or other pumped sources for personnel use
(4) Hydraulic systems (200 lb, 1500 lb, or other pressures as required)
(5) Sewer systems (sanitary, industrial, storm)
(6) Gas systems (natural, propane, oxygen), including compressors
(7) Fire pumps and associated water supplies
(8) Steam systems (low and high pressures)
(9) Ventilation (building and process)
(10) Vacuum systems

(11) Compressed air for instrumentation

Additional systems may exist in some plants, but the list will alert a plant manager or engineer to the various needs and potential losses should electric power not be available.

Systems may be required for manufacturing and services to maintain other services. For example, electric power, water, and compressed air for boilers used to supply steam for the generation of electricity.

A 0.1 to 5 s power failure may cause operating engineers to spend minutes or hours restarting equipment and making adjustments until all systems are again stable. Such disruptions to production should be prevented if economically justifiable.

3.4.3 *Orderly Shutdown of Mechanical Utility Systems.* An orderly shutdown may be acceptable and can be provided with a short-time smaller supply source of power for the following requirements:

(1) To maintain temperature or pressure on vulcanizers until the product can be finished.

(2) To maintain hydraulic pressure until a batch process is completed, or until the pressure can be released without loss.

(3) To operate pumps for a time until all process water has been shut off or has drained back into the sumps to prevent flooding. This becomes serious when the water is contaminated with oil and other waste when flooding occurs and special handling may be needed to comply with antipollution requirements.

(4) To maintain ventilation to clear explosive atmospheres while a normal shutdown proceeds. Purging air is critical to some oven-drying processes.

(5) To prevent sanitary sewers from overflowing before personnel evacuation takes place.

(6) To run gas compressors for the finishing of a critical process.

A complete listing has not been intended, but representative needs of various types should be enough to alert a plant engineer so that a plan may be prepared for an orderly shutdown. Management may then act on the needs justified without being caught unprepared in an emergency situation. Although management may not approve the recommendations immediately, the plan will be ready for resubmitting, approval, and implementation when an interruption occurs, alerting management to the need for an emergency or standby power sypply.

3.4.4 *Alternates to Orderly Shutdown.* An orderly shutdown may not be acceptable. Then an alternate may be selected as follows:

(1) Maintain utilities without an interruption. Full capacity standby power must be available as well as uninterruptible power for boiler controls, on-line computers, and essential relays and motor starters.

(2) Accept an outage but return on standby power. Full capacity power must be available. Startup of functions will be required since magnetic motor starters and relays will have dropped out. Although power is off for 0.1 to 5 s it will take 15 min to ½ h or more for return to normal operation.

3.5 Heating

3.5.1 *Maintaining Steam Production.* Continuous-process plants require uninterrupted steam production. Minimum requirements for continuous steam production are sufficient combustion air, air to instruments and actuators, water and fuel supplies, plus a continuous power supply to most flame supervision systems. The maximum interruption tolerable is that duration during which the inertia of

GENERAL NEED GUIDELINES

Table 4
Systems for Continued Steam Production

Components	Allowable Outage Duration	Systems
Flame supervision systems	Nil	Mechanical stored energy systems Motor generator set ride-through UPS systems
Motor controls and instrumentation	Nil	Same as above
Boiler fans	½ to 2 s	Multiple utility services either on line and relayed, or off line, switched, and transferred
Air compressor	To 30 min, depending on storage; nonessential air users should be automatically shut off	Multiple utility services Turbine- or engine-driven generator, off line Turbine (combustion, steam, water) off line
Water pump	To 5 min, depending on water drum capacity and upsets in steam production caused by power disturbance	Multiple utility services either on line and relayed, or off line, switched, and transferred Turbine (combustion, steam, water) off line, automatic start
Oil pumps (for burners)	To 15 cycles, more with flywheel	Multiple utility services, on line and relayed Turbine (combustion, steam, water) on line
Electric oil supply pumps	Several minutes	Turbine- or engine-driven generator, off line Multiple utility services, on or off line

the fans or pumping equipment will maintain flows and pressures above minimum limits. Table 4 illustrates how this can be achieved.

3.5.2 *Process Heating.* Process heating is defined as heat required to maintain certain process materials at the required temperature. Noncritical heating processes, due to the inherent heat capacities of such systems, can withstand a power interruption of considerable duration, say, 5 min to a maximum of several hours.

Any of the following systems would be adequate for this application:

(1) Engine-driven generators, off line
(2) Multiple utility services on line and relayed, or off line, switches, and transfers
(3) Turbine (combustion, steam, water) off line or on line
(4) Mechanical stored-energy system, with auxiliary motor generator sets off line

Other heating processes such as cord and fabric treating and drying are of such a critical nature that loss of heat will cause an out-of-specification product within 10 s, but the gas or oil burners and flame detectors are sensitive to drops in voltage of about 40 percent for a second or less. In this case, an uninterruptible supply for the controls and an engine generator for a 10 s main power supply or an alternate feed system may be required.

Infrared drying of enamel on automobiles and appliances is a form of heating by electric energy which must be maintained. A short interruption, perhaps 10 s, may be acceptable to bring on-line standby generation or perform automatic switching to an auxiliary power source.

Losses may be substantial should power be lost during the heat treating of metals and when either direct or indirect melting of metals is in process. Two evaluations must be made, one based on not losing the flame on the fuels used, and the other based on accepting an interruption for a short period with a restart necessary. Uninterruptible power is several times more expensive than switching or emergency power, but saves the product and prevents process interruption. Switching or emergency power is less expensive, but product or process losses may exceed the initial savings in the chosen electric supply system.

Induction and dielectric heating are other forms of electrical heating which may or may not allow short interruptions to be tolerated. Most processes, whether in production or in other fields, could tolerate an interruption of sufficient duration to come on line with switched sources or an engine- or turbine-driven generator.

3.5.3 *Building Heating.* Buildings, even in the coldest regions, can usually be without heating for a minimum of 30 min. The same systems listed for noncritical heating processes will be suitable for this application.

3.6 Refrigeration

3.6.1 *Requirements of Selected Refrigeration Applications.* Requirements for refrigeration are usually noncritical for short power interruptions of several minutes to several hours. The need may become extremely critical as the length of time of the outage increases. Consider these refrigeration needs:

(1) Production of ice cream or the freezing of foods may stop in the middle of the process. Not only will all production be lost during a power failure, but damage may result to the product in process.

(2) Material in storage may be in jeopardy as temperatures rise. Cafeterias, frozen food lockers, meat cooling and storage facilities, dairies, and other food operations require refrigeration and will soon be in trouble as the length of the power outage increases.

(3) Scientific tests of long duration may require accurately maintained low temperatures. Short outages of electric power may destroy the tests and require repeating. An expensive and time-consuming process should be provided with a standby power system.

(4) Medical facilities require refrigeration for blood banks, antibiotics, and for long-range laboratory experiments and cultures which could be spoiled.

(5) As the state of the art of superconductivity develops, improves, and spreads to applicable fields, power for cryogenic refrigeration equipment operation will probably become critical.

Where present refrigeration units are electrically driven, when new units of moderate size are to be installed in permanent locations, and when other needs exist (as is usually the case) for emergency or standby power, a common engine-driven generator or alternate utility source should be considered.

Because of the slow rise in temperature of cold-storage facilities, a savings may be made by the use of a smaller than normal standby electric generator. By switching power to various units in turn, an acceptable storage temperature may be maintained until normal electric power has been reestablished.

3.6.2 *Refrigeration to Reduce Hazards.* Certain chemical processes are exothermal and release heat during the chemical reaction. Loss of the cooling or refrigeration system may cause severe damage or even an explosion.

3.6.3 *Typical System to Maintain Refrigeration.* A manual starting of an engine-driven generator, turbine, or alternate utility supply will usually suffice, providing a suitable alarm is installed to notify responsible persons of a loss of refrigeration.

3.7 Production

3.7.1 *Justification for Maintaining Production in an Industrial Facility.* Loss prevention in production facilities due to a power failure is justified on the total sum of many tangible and intangible savings. Some of these items to consider are as follows.

Is there a guaranteed wage clause in the labor contract? If so, there will be a direct loss in wages paid for which no production is received. Where power requirements are low and the heavy power demanding machines are left shut down until normal power is returned, a small electric supply system can be justified to supply finishing areas, inspection areas, office areas and other areas where most of the people work.

Who is waiting for the product? There are periods when the products are being routed to warehouses, and machines are not running at capacity. In this case the cost of machine down time is lower than it would be for a product which a customer will not receive or will receive late if production is at full capacity and an outage occurs.

What is the cost of product spoiled in process? In the rubber industry material may become sealed in vulcanizers, extruders, or mixers at high temperatures and will be difficult and expensive to remove. Steam or water pressure may drop to zero and prevent proper cures with losses due to poor or ruined products.

If all electric power is lost during certain processes in the making of sugar, glass, steel, pharmaceuticals, rubber, paper, chemicals, and some other materials, the product must be scrapped.

What is the cost of consequential damages? While the material ruined may be scrap, there may be as many problems and costs associated with its removal as with the loss itself. Some material must be dug out, or removed by hand, piece by piece, until lines are cleared or chambers are empty and clean so that an orderly startup can follow.

What is lost in reestablishing work efficiency? A ½ h electric power interruption disorganizes the workers. Experience indicates that, following the interruption, it will take men and women at least two or more hours to settle down, go to work, and reach the production level at which they were operating just prior to the power failure. It may take days to reestablish normal procedures in scheduling of incoming and outgoing materials and in telegrams, letters, notices, and calls explaining delays and changing promises.

A less tangible item lost is good will. For example, in the film processing industry the customer may not consider the replacement of the exposed film with unexposed as adequate compensation for his "once in a lifetime" pictures which were spoiled due to a power failure.

Real and potential costs and losses must be calculated or estimated and added together to justify an emergency and standby power system for industrial and commercial facilities.

A reasonable estimate of the costs associated with each past power failure should be calculated and recorded in a journal with the date, duration, and conditions existing at the time. As time goes on this will be found to be valuable factual backup information for budget requests.

3.7.2 *Equations for Determining Cost of Power Interruptions.* A rough estimate of the cost of a power failure from a cash flow viewpoint may be calculated as follows:

Total cost of a power failure $= E + H + I$

Where

E = Cost of labor for employees affected, in dollars.

H = Scrap loss due to power failure, in dollars

I = Cost of startup, in dollars

The value of E, H, and I may be calculated as follows:

$E = AD\,(1.5B + C)$
$H = FG$
$I = JK\,(B + C) + LG$

Where

A = Number of productive employees affected

B = Base hourly rate of employees affected, in dollars

C = Fringe and overhead hourly cost per employee affected, in dollars

D = Duration of power interruption, in hours

F = Units of scrap material due to power failure

G = Cost per unit of scrap material due to power failure, in dollars

J = Startup time, in hours

K = Number of employees involved in startup

L = Units of scrap material due to startup

After the cost of down time has been calculated the savings in utilities should be subtracted to arrive at a total cost of down time.

3.7.3 *Commercial Buildings.* For commercial establishments a similar example may be assembled based on the length of the power interruption, labor cost, loss of profit on sales, loss due to theft, and startup costs.

3.7.4 *Additional Losses Due to Power Interruptions.* In addition to losses relating to cash flow are those more difficult to calculate but which should be included when available and applicable, such as:

(1) Prorated depreciation of capital costs

(2) Depreciation in quality in process materials

(3) "Cost" of money invested in unused materials or machines

Other losses may occur under special or unusual conditions. In an industrial plant operating at 100 percent capacity any loss in production results in the loss of the profit of the item or service. The prorated cost of fixed and variable overhead becomes a loss. Customers may switch to competitors. Expenditures for standby power have additional justification under this condition.

3.7.5 *Determining Likelihood of Power Failures.* Next must be determined the likelihood of a power failure by studying the record of the plant or utility company electrical supply, or by transmitting the service requirements to the local utility and obtaining their recommendations. Examples of recorded power failures are shown in Table 5.

Rather than complete power failures as recorded in Table 5, Table 6 covers short-term dips.

A projection should be made, working with the utility company, as to whether the power reliability will improve or decline. Since the cost of a power failure, as

Table 5
Example of Recorded Power Failures

Date	Time	Duration	Transmission Line
9 March	09:52	10 min	14
11 June	21:53	12 s	14
11 June	22:13	9 s	14
15 July	20:40	5.5 s	13 +22
17 July	19:13 20:44	1–2 min	14 (9 times)

defined in this publication, is paid by the user, it is important that he relate the reliability of power duration and quality to the need and justification.

3.7.6 *Factors that Increase Likelihood of Power Failures.* As full designed load is reached or exceeded, the probability of a power failure increases. A similar probability exists as systems become more complex and as system equipment becomes older.

3.7.7 *Power Reserves.* Power reserves in the user's area should be investigated. Adequate reserve margins above peakload demands provide a guide to power reliability because the margin provides for some contingencies.

3.7.8 *Examples of Standby Power Applications for Production.* The following typical examples were taken at random of users who found, after evaluation, purchase, and installation, that a standby power system was justified.

(1) A drug company installed a special engine generator backup system based on the need for constant temperatures in manufacturing processes. The value of processes saved during blackouts soon exceeded the cost of the equipment installed by more than 10 times.

(2) A photographic film-processing company specified a 45 kW engine generator when the film-processing machines were ordered. Film must be moved along in each aspect of the process within 30 s or loss of quality will result.

(3) A 400 kW, 480V, three-phase engine generator was installed in a highly automated egg farm. Electric service continuity was mandatory for egg production and the welfare of the "machines" (chickens). During the first year two electric service outages occurred and the automatic emergency generator reliably supplied power.

3.7.9 *Types of Systems to Consider.* An engine- or turbine-driven generator or an alternate independent utility source usually will fulfill the requirements for standby power. More than a standby system may be required for critical loads.

Table 6
Example of Recorded Short-Term Dips

Date	Time	Line	Duration (cycles)	Line-to-Ground Dip (per unit)			Voltage After
				E_a	E_b	E_c	
14 April	21:50	32	18	0.86	0.81	0.75	1.0
30 April	15:53	System	43	0.83	0.92	0.92	1.0
9 May	07:52	23,24	32	0.90	0.85	0.83	1.0
9 May	07:54	20	24	0.31	0.31	0.58	1.0
9 May	07:55	22	46	1.00	1.00	0.69 0.40 0.50 0.06	1.0
17 May	00:43	13,22	21	0.50	0.69	0.31	1.0

Motor starters, contactors, and relays held closed by a coil and magnetic structure are especially sensitive to short-time power outages or voltage dips. Their drop-out characteristics vary with respect to voltage level and the length of duration of the voltage dip. As a guide, a voltage dip to 70 or 60 percent of rated voltage for 0.5 s will deenergize many of the devices. The longer the time of the voltage dip, the more devices will deenergize. Depending on the application of these devices, an emergency or uninterruptible power supply system may be justifiable, especially where boiler controls, critical chemical processes, safety devices, and other critical systems are required to be maintained [4].

Devices much more likely to fail or malfunction due to voltage excursions are modern static switching devices, which have almost no time delay, and pressure, temperature, and flow transmitters and receivers of the electronic type. Automatic calculators for process control, data reduction, and logging are in this same category [5].

An uninterruptible power supply system for critical on-line computer or boiler control loads may be required since a 30 cycle outage may mean restarting the plant. In combination the uninterruptible power supply system may be of short duration since the standby equipment should be designed to be on line in from 10 to 60 s.

Factory clocks used for production control, or for the basis of incentive wage payments, should be arranged to maintain accurate time by the use of batteries for power during the time of any power interruption.

Support of production facilities will justify most or all of the user's needs detailed in Section 3. The evaluation, justification, and decision to purchase and install a standby, emergency, or uninterruptible power supply system, or a combination of these systems, must include the consideration of all the electric power requirements for all listed needs in case of a power failure.

3.8 Space Conditioning

3.8.1 *Definition.* Space conditioning is a controlled environment, either to maintain standard ambient conditions or some artificial alteration of a standard environment in a building, room, or other enclosure.

3.8.2 *Description.* A controlled environment may include any of the following variables:
 (1) Temperature
 (2) Vapor content
 (3) Ventilation
 (4) Lighting
 (5) Sound
 (6) Odor
 (7) Gas
 (8) Dust
 (9) Organisms

3.8.3 *Codes and Standards.* Codes apply primarily where personnel safety is involved, but still very little reference is made specifically to power supply requirements. Most requirements in OSHA are dedicated to ventilation of areas where hazardous gas, contaminants, and other similar elements and conditions can exist that would endanger personnel safety. Sections 31.5518, 1910.1006, and 1910.96 of the 1975 National Electrical Code give some such requirements. Where "adequate" ventilation is referenced, conditions exist which may warrant backup or uninterruptible power supplies to provide this ventilation.

3.8.4 *Application Considerations.* Air-conditioning loads for personal comfort are not normally considered critical and are even shed under overload conditions in some cases. Where equipment is sensitive to temperature, however, such as where solid-state electronic components exist, air conditioning can be critical. Where continuous backup or standby power is not available, a self-contained emergency or standby generator would apply. An interruptible power supply specifically for this purpose is not normally necessary since loss of power would cause no instantaneous temperature change. Likewise, where moisture and humidity cause serious equipment and operating problems, heating might require some type of backup power supply.

Table 2 shows examples of tolerable outage time for various applications. Ideally, tests should be run by the designer or planner to determine the maximum transfer times tolerable. Amount of power needed, acceptable outage time, and economics will determine the applicable power supply. Often economics dictate that power for space conditioning be incidental to total power supply requirements, and the user must evaluate the ultimate consequences of loss of power. An adequate alarm system might be a consideration in this case.

Alarm and signal circuits often require as much or more reliability than equipment providing critical space conditioning. An uninterruptible power supply would be necessary where even a temporary loss of power would endanger personnel or cause severe equipment damage.

3.8.5 *Examples of Space Conditioning Where Auxiliary Power May Be Justified.* Typical situations and facilities in which a comprehensive study of the need for supplemental electric power is warranted are the following.

(1) Commercial and laboratory horticultural botanical installations may require programmed cyclic control of temperature, humidity, and lights to develop the crop yield or desired experimental results. The loss of temperature or humidity control for 6 to 8 h can result in total loss of a crop. A greenhouse must have maintained temperature in order to produce.

(2) Tropical animal raising requires control of ventilation, temperature, humidity, and lighting. All are completely dependent on electric power. A loss of heat or cooling can result in death or illness to all animals being raised. Lighting and temperature changes from the established cycle can induce unwanted breeding periods in many exotic creatures. Egg production may be greatly curtailed by temperature changes or loss of light.

(3) Agricultural operations are often located in remote areas where the long utility lines are susceptible to damage.

(4) Final operations and packaging of materials susceptible to contamination are conducted in "clean room" type environments. A power interruption will shut down the total operation, and contamination may result as people exit, unless the room is kept under positive pressure to prevent in-drafts from bringing in contaminants. Such contaminants will necessitate a complete recleaning of the room before it can be used again.

(5) In large processing plants, priorities may be established for essential loads. Nonessential loads may be dropped automatically by load-shedding systems in the event of a power failure, and other limited emergency sources must be relied upon. A reevaluation of these priorities should be undertaken. Critical temperature controls may have been placed in an air-conditioned space. These controls must function to achieve an orderly process shutdown, but can fail due to overheating caused by the lack of ventilation or cooling.

(6) Windowless buildings or inside

rooms may become unsafe for occupancy during an extended power interruption. Power supplies to keep these areas ventilated should be installed if evacuation of all personnel is not acceptable.

3.8.6 *Typical Auxiliary Power Systems.* Needs may require any one or more types of emergency and standby power systems available. An engine-driven generator or combustion turbine should be considered for these applications. Switching to a separate alternate electric utility line would involve a lower capital cost, should such a line be readily available. In addition, a short-time uninterruptible power supply system may be required for critical applications to supply power during engine startup or switching.

3.9 Fire Protection

3.9.1 *Codes, Rules, and Regulations.* Various codes, standards, laws, rules, and regulations contain either advisory or mandatory statements related to emergency and standby power systems for fire protection. See Title 24, California Administrative Code, Part 3, Basic Electrical Regulations (Article E700, Emergency Systems); NFPA No 101-1973, Life Safety Code; IEEE Committee Report [6], pp 19, 20; and Katz [7]. There appears to be more written concerning the wiring system reliability and needs than the source of electric power supply or the checking and maintenance of the complete installation. All three are vital.

Article 760 of the 1978 National Electrical Code (NEC) lists some requirements for protecting circuitry in alarm and signal systems. Article 230-94 pertains more to the power supply by allowing a fire-alarm circuit to be connected to the supply side of a service overcurrent device if it has separate overcurrent protection.

The requirements of all local, state, and national standards and codes should be determined. The insurance company who will underwrite the insurance can provide valuable assistance in making sure that all requirements are met.

A common-sense approach should be used even beyond meeting the letter of the law and insurance requirements as a minimum standard. The real goal is to avoid a destructive fire, or in the event a fire does start, that it be held to a local area with minimum damage to property and no harm to personnel. Knowledge in depth of the industrial facility and processes by plant engineers and other responsible plant personnel should be utilized to reduce the likelihood of a fire and the extent of damage should one begin.

3.9.2 *Arson.* Arson may be a source of a fire and the need of power for plant security, lighting, signaling, and communication covered in other sections should be regarded as contributing to the reduction of possible loss by fire from this cause.

3.9.3 *Typical Needs.* The user's possible specific needs connected with the general need of emergency and standby power systems for fire protection include the following:

(1) Power, usually batteries, to crank the engine on an engine-driven fire pump

(2) Sprinkler-flow alarm systems

(3) Communication power to notify the fire department and to assist in guiding their activities

(4) Lights for the firemen to work by in the buildings, around the outside area, and mobile on company trucks

(5) Power for the boilers which supply steam-driven fire pumps

(6) Motors driving fire pumps, well pumps, and booster pumps

(7) Air compressors associated with fire water tanks

(8) Smoke and heat alarms

(9) Deluge valves
(10) Electrically operated plant gates, drawbridges, etc
(11) Communications such as public-address systems for directing evacuation of personnel
(12) Fire and hazardous gas detectors

3.9.4 *Application Considerations.* A fire almost always warrants initiation of an emergency shutdown in a plant either by operator-initiated devices or by automatic operation. The circuit required for shutdown is obviously critical as are all main circuits for the abovementioned fire protection equipment. Uninterruptible power supplies should be first choice for these applications. Where automatic fire protection is employed, like sprinkler systems, CO_2 discharge, etc, nuisance initiations must be prevented, a condition which lends itself to application of uninterruptible power supplies.

It is common practice in large plants to back up an electrically driven fire water pump with a mechanical driven one such as a diesel drive. If all ac power is lost in an emergency shutdown operation, fire protection is maintained. In this case, the ac power supply to the motor-driven pump is not absolutely critical. If a mechanical drive pump is not available, the motor-driven pump would normally be supplied by an emergency generator.

Under emergency conditions, such as fire, communications are sometimes vital to personnel and equipment safety. Under emergency conditions, communications should not be subjected to an outage of any kind. Even momentary outages of a few cycles might cause erroneous communications, especially where remote supervisory control is employed.

3.9.5 *Feeder Routing to Fire Protection Equipment.* Electric power distribution systems supplying fire equipment should be routed so as not to be burned out by a fire in the area they are protecting.

Protection of control circuitry as well as main fire protection circuitry can be enhanced by underground conduits, separate conduit and wire from other circuits, and fire-resistant insulated cable.

3.10 Data Processing

3.10.1 *Classification of Systems.* Most data processing installations can be grouped into two general classes of operation in accordance with their usage. These classifications are off line and on line. These categories will be helpful in identifying a data processing system's vulnerability to electrical power disturbances, since an off line process will rarely require power buffering or backup sources or equipment. Conversely, it is common for an on line system to warrant the additional expense of buffering or backup equipment.

Off line data processing systems are generally set up to perform one or more programs at a time in a sequential or batch type. Usually such systems have a program operator for automatic processing of control cards for a given job run. Often, 24 h operation for a heavily loaded system may be necessary. In the off line category, for the most part, are business, scientific, and computer center applications. Systems of this type are particularly vulnerable when the programs are lengthy (several hours in duration). Thus the insertion of several natural breaks or checkpoints in the program for segmentation of long programs is highly desirable. Programming can be designed to save intermediate results at a checkpoint and to have the option of restarting at the last checkpoint which preceded the power interruption. Such a practice in program interruption can be valuable in protecting against peripheral equipment failure. Many current programs are being designed without checkpoint techniques even though the practice has been found to be feasible. Data dependent programs have running times which vary in duration of the magnitude of input data, and accordingly it is difficult to limit the run to much less than a 20 min period.

On line systems, or as they are often called, "real time" systems, are systems which are time and event oriented. They must respond to events which occur randomly in time, often coincidentally. An awareness by the system of events which occur that are external to the computer and beyond its influence is a requirement. In this category are such applications as industrial process monitoring and control systems, air line passenger reservations systems, vehicular traffic control, certain specialized scheduling applications, international credit card and bank associated credit/transaction systems, plus many more. With these systems the computer outage problem due to power interruption is usually more critical than in off line applications. Further, there is generally no merit in segmenting programs. In most cases any outage or power interruption will result in the loss of some data which was available only during the time period of the outage. The form of the input data is not conveniently available for a rerun, but may come from sensors such as thermocouples or pressure transducers which are scanned by a computer. When a computer controls a process, potential problems resulting from a power disturbance are generally serious enough in terms of product damage or equipment malfunctioning to warrant the use of a reserve or backup power source. Further, any solution, to be adequate, must accomplish the necessary switching to the backup source without power interruption to the computer. It is obvious that the potential losses to several hundred users, or input stations, to a time-shared computer system would warrant the providing of a backup source which can practically guarantee uninterruptible power. In cases where equipment or process monitoring must be made, as through data logging, protection from destruction of only the core memory content during a power interruption may be adequate. Automatic restart upon return of power is possible to minimize time that the equipment is down and can often be utilized with the additional provision for a manual means of updating the system data or information not gathered or scanned during an interruption.

The foregoing discussion has categorized data processing equipment and resulting electrical loads by their respective functional usage, that is, whether they are off line or on line systems. In most cases, large systems are involved. A further differentiation which can be made is by the magnitude of load. As with most power utilization equipment, smaller power consuming devices can generally be supplied from a single-phase source. The differentiation between single-phase versus three-phase power consuming systems is often necessary since the methods of protecting against input power disturbances and outages can be quite different for each system. Some of the data processing systems which use single-phase power will employ microprocessors or minicomputers. Others may consist of multiple single-phase load units distributed in their connection to three-phase power so as to achieve a reasonable load balance when all units are operating. This may result in load unbalance when some of the units are turned off.

In general, computers and peripheral units which draw less than 1.5 kVA will often be single phase. Those which draw more than 10 kVA often require three-phase power. In most cases single-phase loads can be connected to three-phase sources provided load unbalance at maximum load is not excessive, generally taken as 25 percent or less.

3.10.2 *Needs of Data Processing Equipment from a User's Viewpoint.* In general, data are gathered in analog or digital form and converted to one or the other system. Refer to definitions in the preceding paragraphs. These data in electrical form are then processed through one or several programs by an electronic computer, minicomputer, or microprocessor. The results are traced or printed out, or a signal is generated and fed back to form a closed-loop system which provides control to match preset conditions. Both types of outputs are commonly available and used.

Both industrial and commercial users apply real-time data processing systems with a computer on line. Failure of the data processing system often causes loss of valuable data or interruption of a critical process, either of which may result in extensive financial losses and require hours, days, or even weeks to fully recover. Failure of computers providing control may also jeopardize the safety of personnel. These failures can occur as a result of an electric power failure. Thus a reliable, high-quality primary power supply is frequently justified to minimize loss of money and to prevent injury or loss of life.

A short list of control operations which frequently include data processing equipment, minicomputers, or microprocessors will suffice to alert users to the types of hazards and losses that may occur as a result of a power failure. Applications are categorized for both industrial and commercial classifications.

(1) Industrial applications: materials handling; regulating control and status monitoring of pipelines, compressor, and pumping stations at remote locations; mixing compounds; grinding, drilling, machining; twisting textile fibers; steel mill processing; refining (oil); automatic testing and gauging; fabric processing; power flow and load dispatching; safety and security monitoring; process control and data acquisition

(2) Commercial applications: controlling elevators; automated checkstand combined with inventory control; newspaper production machines; airline passenger reservations; computer-aided emergency vehicle dispatching; environmental, life safety, security monitoring and control for buildings; typesetting; accounting; traffic control—cars, rail, and airplane; stock exchanges and broker transactions; corporation computers for engineering, scientific, and business matters; hospital diagnostic equipment and individual intensive care systems; banking, financial, and credit transactions

With the advent of the electronic computer as a part of data processing and process control, an increased emphasis has been placed on the need for emergency and standby power systems to assure a continuous flow of energy. Coupled with this need is a superimposed problem, namely, the suppression of most short-time switching interruptions, voltage surges, dips, and frequency excursions. Often the suppression of transient disturbances and the need for emergency or standby power can be satisfied through a single installation of supplementary or auxiliary equipment. These transient disturbances have been a part of the electric power supply in the past, but caused few problems until electronic equipment came into extensive use. In some instances the solid-state power control equipment has caused the problems particularly on small, independent power systems.

3.10.3 *Power Requirements for Data Processing Equipment.* By the proper selection of an electric supply system, the power needs associated with data processing with a computer can be met, namely, a reliable source of noise-free electric power at all times and of a much higher quality than previously demanded by most devices.

The problem is how to reconcile commercial short-duration power interruptions with relatively short time domains in electronic circuits. In early stages of data processing equipment and computer equipment development, it was not unusual to experience problems with hardware and software when power disturbances of microseconds in duration were experienced. Most equipment built in that era was extremely vulnerable to such short time disturbances. The goal of most manufacturers in today's technology is to build from 4 ms to 1 cycle of carryover, or ride-through time, into their equipment.

Table 7 shows computer input power quality parameters for several manufacturers. The user should consider Table 7 only as a source of some examples since computer designs vary with size of computers, their processing power, and the technology available when the design was created. They are continually changing and the parameters of power needs are changing rapidly with the designs. Some of the paragraphs which follow expand on the individual parameters which are presented in the table. Although there is a degree of variance between computer manufacturers, the following represents the principal power parameters which are considered important by most major companies. While several of these parameters, such as frequency variation, can be

Table 7
Typical Range of Input Power Quality and Load Parameters of Major Computer Manufacturers

Parameters*	Range or Maximum
1) Voltage regulation, steady state	+5, −10 to +10%, −15% (ANSI C84.1−1970 is +6, −13%
2) Voltage disturbances	
Momentary undervoltage	−25 to −30% for less than 0.5 s with −100% acceptable for 4 to 20 ms
Transient overvoltage	+150 to 200% for less than 0.2 ms
3) Voltage harmonic distortion**	3−5% (with linear load)
4) Noise	No standard
5) Frequency variation	60 Hz ± 0.5 Hz to ± 1 Hz
6) Frequency rate of change	1 Hz/s (slew rate)
7) 3ϕ, Phase voltage unbalance***	2.5 to 5%
8) 3ϕ Load unbalance****	5 to 20% maximum for any one phase
9) Power factor	0.8 to 0.9
10) Load demand	0.75 to 0.85 (of connected load)

  * Parameters 1), 2), 5), and 6) depend on the power source while parameters 3), 4), and 7) are the product of an interaction of source and load and parameters 8), 9), and 10) depend on the computer load alone.
  ** Computed as the sum of all harmonic voltages added vectorially.
  *** Computed as follows:
$$\% \text{ phase voltage unbalance} = \frac{3(V_{\max} - V_{\min})}{Va + Vb + Vc} \times 100$$
**** Computed as difference from average single-phase load.

relatively insignificant when power is derived from a commercial power source which embodies vast tie networks, they can become an important design consideration when supplemental or independent power sources are applied as a means of power quality improvement.

208Y/120 V single- and three-phase voltage is the most common computer unit utilization voltage with some single-phase 120, 120/240, or 240 V. Some equipment is reconnectable for use at several voltages by use of an internal tapped transformer. Tolerance on the 60 Hz voltage varies between manufacturers; however, limits as listed in ANSI C84.1-1977, Voltage Ratings for Electric Power Systems and Equipment (60 Hz), are +6 and −13 percent.

Systems requiring 400 Hz power (for mainframe and some peripheral devices) often derive their power from a motor-generator set which is usually an integral part of the computer system. The motor of these units, ranging from 10 to 20 horsepower, is ordinarily connected to a power source which is separated or isolated from the computer equipment. A common nominal line voltage is 480 V, three phase, for which the rated motor-generator input voltage is 460 V ± 10 percent. The output voltage of the generator is closely regulated, usually to within ± 2 percent, and is distributed over special 400 Hz lines to the load. Smaller computers and peripherals may obtain power from within the computer unit or a nearby unit from small inverters within the equipment. These units are usually not greater than 20 kVA and are powered from the input 208Y/120 V three-phase service or a 120/240 V single-phase device. Since inverter characteristics during abnormal operations, such as inverter failure, are quite different from motor-generator set failure characteristics, it would be well to check with the computer equipment manufacturer before investing in supply equipment.

A growing trend is to provide regulated low voltage direct current for logic circuits by rectifying the output from 400 to 40 000 Hz static inverters. These devices, though lacking ride-through ability when installed without the addition of sizable banks of energy storage capacitors, do offer a compact and convenient source of regulated low voltage dc power. With proper design and redundancy, their reliability is at least equal to that of a central motor generator.

As discussed later, where 400 Hz or other special frequency power must be distributed to computer equipment from a location remote from the computer equipment, special considerations must be made for voltage drop (increases with higher frequency systems). Conductors are ordinarily installed in nonferrous raceways.

Computer manufacturers usually specify maximum momentary voltage deviations within which their equipment can operate without sustaining errors or equipment damage. The transient conditions are defined in terms of amplitude and time duration. Historically, an example was a range from ± 5 percent of undefined duration to −30 percent for 500 ms, and +130 percent for 5 ms. Another example is ± 20 percent for 30 ms. A few manufacturers also specify a duration limit for total voltage loss of from 1 ms to 1 cycle. A figure of 8.3 ms (½ cycle) for older equipment was typical. These values should not be confused with impulse tolerances, which are of much shorter duration (microseconds) and higher level (500 percent) and are usually part of the noise susceptibility and electromagnetic compatibility (EMC) tests.

Fig 4 shows an envelope of voltage tol-

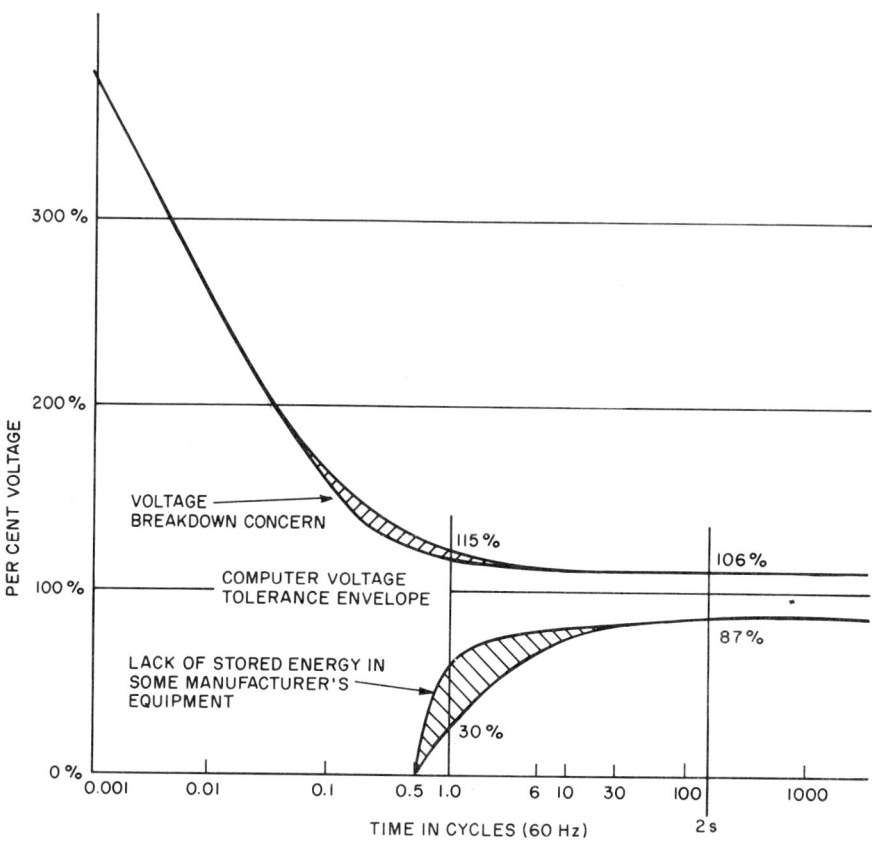

**Fig 4
Typical Design Goals of
Power-Conscious Computer Manufacturers**

erances which is representative of the present design goal of a cross section of the electronic equipment manufacturing industry. Shorter duration overvoltages have higher voltage limits. Some computer manufacturers specify a maximum allowable limit for volt-seconds, typically 130 percent of nominal volt-seconds (area under the sine wave).

The manufacturer's tolerance on 60 Hz equipment ranges from ± 0.5 Hz to ± 1 percent, with the majority of the equipments limited to ± 0.5 Hz. Time-related peripheral devices are most sensitive to frequency (clocks, card readers, magnetic tapes, discs). Ferroresonant (ac-dc) supplies, which are widely used by some manufacturers in peripherals, are also frequency sensitive since they operate on a tuned circuit principle. They can generally tolerate variations of only ± 1 percent. Some auxiliary motor generators and other power supplies can tolerate variations as wide as ± 3 Hz; however, for satisfactory operation of the entire system, the ± 0.5 Hz tolerance should be maintained at all times for all parts of the system. Deviation from the tolerance may cause equipment malfunction or damage. 400 or 415 Hz power is used primarily in computer mainframes or in other areas where high density power is needed. The higher frequency allows design of a smaller and more compact power supply and components which result in reduced heating losses. The deriving of 400 Hz power does involve additional costs. Also, distribution of 400 Hz power presents some special problems. Thus for many of the peripheral manufacturers, 60 Hz power is still preferred. Some equipment, such as large CPUs, use both 400 and 60 Hz power, with 60 Hz mainly for cooling fans and blowers. The current trend is to use 400 Hz power in the larger data processing systems, with 60 Hz limited to the smaller systems and peripherals. Development and wider usage of 60 to 400 Hz or higher frequency static inverters to replace motor-generator sets and to be placed in individual equipments would reverse the trend back toward 60 Hz input in equipments. Several major computer hardware manufacturers predict the use of frequencies considerably above 400 Hz for future generation computer systems. Inverter frequencies of 20 kHZ are commonly used now and some state-of-the-art inverters operate at 100 kHZ and higher.

The voltage quality and characteristics of 400 Hz power are extremely critical with frequency generally not as critical a parameter. Most manufacturers, who consider the 400 Hz source an integral part of the computer, would prefer to furnish conversion equipment as part of the computer installation. Further, because of the special problems associated with the distribution of 400 Hz power, conversion equipment is generally located relatively close to the utilization equipment.

In addition to being sensitive to the limits of $\pm$ 0.5 Hz, some system peripherals are also sensitive to the rate of change within this band. Though extensive information is not available, a variation of 0.05 Hz/s has been cited by one manufacturer as a limit for some units of his system. A typical limit is 1.5 Hz/s, measured as rate of change in a 10 c running average. The limit is most significant when turbine or engine generators are applied where small load step applications may cause the rate to be exceeded.

The maximum harmonic distortion permitted on input lines ranges from 3 to 5 percent, with the majority at + 5 percent. The percentage is usually specified as total line-to-line distortion, with a maximum of 3 percent for any one harmonic.

Excessive harmonic content can cause heating in magnetic (iron) devices such as transformers, motors, and chokes. The harmonic distortion will also appear as additional ripple in the output of some ac-dc power supplies and also cause threshold limits to vary in peak and average sensing circuits. Either can contribute to data errors.

It should be noted that elements of the load may introduce considerable distortion or "noise" into the power source. This reflected "noise," though not in the source, may require suppression through filtering to avoid interference with other loads also connected to the system.

Though some manufacturers build only single-phase equipment, the majority of equipment, particularly for larger systems, is three-phase. The maximum deviation from normal 120° spacing ranges from $\pm$ 2.5° to $\pm$ 6°. Unequal phase displacement, whether in the source or due to unequal loading, can further contribute to phase voltage unbalance.

Though not specified by all manufacturers, the maximum phase-to-phase voltage unbalance (defined below), with a balanced three-phase load, should be in the range of 3 percent. Unequal distribution of single phase, as is frequently encountered in computer systems, will increase the amount of voltage unbalance. Wherever possible, effort should be made to distribute load evenly between the phases.

Excessive phase voltage unbalance can cause excessive heating to three-phase devices such as motors. Similarly, relays and other electromechanical devices may be damaged due to continuous operation at high (or low) voltage. In addition, high "ripple" may be observed in some three-phase ac-dc power supplies if the voltage unbalance to the supply is high. Percent voltage unbalance is defined as

$$\frac{3(V_{max} - V_{min})}{Va + Vb + Vc} \times 100$$

The voltage envelope should be sinusoidal, with a crest factor of 1.414 ± 0.1. Waveform deviation should be limited to ± 10 percent line to neutral. The variation in the amplitude (in time) of the wave should not exceed ± ½ percent.

Excessive modulation of the voltage can produce pulsing and speed variations in motors and can introduce additional ripple in the output of ac-dc power supplies.

Some computer units have half-wave rectifier units and SCRs (half-wave phase control). These are capable of creating a dc component of load current and greater current in the neutral than in the phase conductors. Power sources for such loads must be capable of handling them.

It is well known that electronic equipment is capable of reaching design performance levels only when there is proper cooling introduced. The principal thrust of this discussion involves the electrical power needs of the data processing equipment. However, it must be recognized that cooling to that equipment can be an overriding and sometimes limiting factor in the operation of data processing equipment, principally because most cooling and ventilating equipment is electrically powered. In addition to comfort air conditioning for the computer room, the computer (CPU) logic and chassis often have special cooling requirements, and many have overtemperature alarms or cutoffs. With larger systems and increased capabilities computers have, for technical reasons, also grown more compact, leading to increased power densities in terms of watts per cubic foot. The result has been a progression in chassis cooling from natural convection, to forced air, to chilled water, and in some cases, to a refrigerant-cooled chassis. One manufacturer uses a dual freon unit for cooling the mainframe alone. However, intermediate-size computer installations are frequently cooled by forced air through ducts or underfloor areas. Small computers and peripherals may contain their own internal fans and draw cooling air from the room where they are installed.

The methods of cooling of a computer system can also heavily influence how long a computer system can operate without electrical power to cooling apparatus. Assume that computer power could be maintained through the use of batteries, standby generators, or other auxiliary equipment, which is described in later paragraphs. It then becomes important to consider the design/selection of computer system cooling systems and their need for auxiliary power supplies. Obviously, there is no need to provide battery ampere-hour capacity, which can extend computer operating time beyond the time which a computer system can operate before it must shut down due to overheating.

Most manufacturers are reluctant to cite these times for their computer hardware since many variables within a facility can exist. For forced-air cooling, many claim up to 15 min can elapse before overheating and equipment shutdown occurs due to operation of overtemperature sensors. For those systems which use chilled-water cooling, approximately 2 to 3 min has been given as the length of time that chilled supply can be cut off. For the refrigerant cooling, about 15 min is often considered the maximum.

Most larger systems shut down automatically when a high temperature is reached and many provide a warning alarm when approaching a high temperature.

The need for, and extent of, introduction of humidification into a data processing equipment room will vary considerably with the geographic area. Humidity control is required to ensure orderly movement of papers, cards, and magnetic tapes in the operation of the computer and its peripherals. Low humidity allows the building up of static electrical charges, which in turn causes cards and tape to stick together, jam, etc. Extremely high humidity can result in condensation of moisture on chilled chassis plates, resulting in rust and corrosion. Most data processing equipment manufacturers recommend that a humidity range of 40 to 60 percent be maintained in equipment rooms. As with cooling apparatus, discussed in the preceding paragraph, humidification generating equipment can be as important as cooling equipment. Any reserve or standby equipment must be sized for not only the supply of electrical power to data processing equipment, but also for the length of time the equipment can operate without supplemental humidification or dehumidification and attendant problems.

3.10.4 *Influence Factors of Data Processing Systems on Incoming or Supplementary Independent Power Sources.* Much of the foregoing discussion has been dedicated to the various influences that the source of electrical power has on data processing equipment and systems. It should be recognized that the same equipment and systems can, in various ways, interact with and influence the source power system and other utilization equipment served by it. Some of these influences become important if supplementary independent power sources are used as an enhancement of power source quality, particularly where emergency and standby power is applied.

The electrical load of a data processing system depends mostly on the makeup, complexity, and even function of the system. A typical small system may vary from 10 to 50 kVA, and larger systems from 100 to 300 kVA. Some multiple systems may be as large as 2000 kVA. As mentioned in the foregoing paragraphs, as systems become larger, some manufacturers elect to power the mainframes with 400 Hz power derived from the 60 Hz source, often through separate motor-generator sets.

A principal computer manufacturer states that approximately 30 to 35 W per square foot per computer floor space should be allowed in planning for ultimate computer electrical power consumption. This load density is exclusive of any attendant space conditioning, humidification, or dehumidification loads which would be required.

For an established system, most load growth will result from the incremental and often unpredictable addition or expansion of peripheral devices (tapes, discs, cardpunch, etc). However, not all system growth results in electrical load growth. For example, the replacement of several smaller discs by a larger (in a computing sense) disc device, may actually require less power. Major load growth occurs when an entire system performance capability is upgraded, which usually results in a new mainframe and some peripherals.

The characteristic power factor of a computer system is relatively high. The power factor of the 60 Hz portion of the load generally ranges from 80 to 85 percent, and for the 400 Hz (motor-generator set) load generally ranges around 90 percent. Depending on the amount of 60 Hz load, and if a motor-generator set is used, the overall combined power factor is usually in the range of 80 to 90 percent. During initial powering up or startup, the power factor may drop as low as 50 percent for short periods.

Most equipment manufacturers attempt to balance their equipment load in the design of the equipment and in load connection in the planning phase. However, load unbalance may run from 5 to 30 percent (phase-to-phase) steady state and up to 100 percent dynamic, as in startup. The effect of load unbalance is to produce unbalanced phase voltages. See the discussion of voltage balance in the preceding paragraphs.

The powering up of a computer system may place severe demands on the power source. In that regard independent power sources are more vulnerable than a commercial power source from the utility company. Efforts are made by manufacturers to reduce inrush by various methods. Energizing of large loads is often sequenced in steps, manually or automatically. Special motor starting techniques can be used to reduce inrush in the starting of large motors, discs, and motor-generator sets.

The user often can employ operational procedures designed to ease startup and step loading. As an example, large groups of peripherals may be started manually, in sequence, rather than simultaneously. Large 400 Hz loads, such as mainframes powered by motor-generator sets, may be brought on line slowly by controlled buildup of the generator output to give the logic load a cushioned start as well as reduced inrush.

Even with reduction methods, high inrush currents are common with many pieces of computer equipment. One manufacturer's central processor whose steady-state load is 24 kVA presents a 1500 percent (of steady state) transient for approximately 100 ms, which decreases to 600 percent in 300 ms. Another manufacturer, requiring 60 kVA steady state, presents a 400 percent transient. Still another manufacturer requiring about 8 kVA steady state presents a 1000 percent transient for ½ cycle, dropping to normal not later than the next cycle. The startup of a 40 kVA motor-generator set can require up to 200 percent for 2 to 10 s, even with special starting methods. High inrush puts an added requirement on the design of the power source, especially on devices such as ride-through motor-generator sets and static inverters. Current limiting protection for inverters is typically 125 to 175 percent. Voltage output will drop during current limiting output.

Even when a system is on line and drawing steady-state power, it may present severe load transients due to the operational demands of the computer. Step changes in load as high as 200 to 300 percent may be possible when starting an additional unit. Frequently, step changes in load can be minimized within the equipment through judicious programming that avoids simultaneous energizing or operation of several processes or pieces of equipment.

Pulsing loads can cause problems and occur when a number of devices that have power peaks, which are coincident and repetitive, are connected to the same power source. As an example, the programming of multiple tape units to rewind simultaneously should be avoided. Line printers whose steady-state load is small may step 100 to 200 percent when striking a full line of print. The cumulative effect of a large number of printers in synchronism can place a severe strain on the power system. If the power source includes a rotating device with speed control, it is possible for the load to cause oscillations in the control and output. Though extremely rare, the problem should be considered if there could be a large number of synchronized pulsing loads.

Certain elements in the load, such as saturated magnetic circuits (transformers, motors), may cause distortion of the voltage waveform. Problems often result from the reflection into the source of load-generated noise, such as switching spikes caused by turning devices on or off or by the firing of high-speed solid-state devices (SCRs, diodes) which are a part of the computer load. The spikes, of microsecond width, may run several hundred volts on the 120 V line. These disturbances may have to be eliminated by filtering to avoid interference with other parts of the load.

The load factor for an on line system may approach 100 percent due to the continuous nature of its operation. Off line systems will generally be lower, depending on their daily scheduling and use.

Demand factor is the ratio of the actual steady-state running load to which a power system will be subjected to the total connected load.

Demand factor is important since connected kVA loads are generally available from the equipment manufacturer as nameplate ratings of individual components and can be taken as the worst case condition. Thus, an arithmetic total of the loads can be computed. However, some computer system manufacturers provide data on the actual running loads, as seen by the total system, when operating the individual components as a system.

Through experience and data obtained from several of the computer systems manufacturers, some of the larger computer systems having connected loads in the area of 300 kVA, experience demand factors of between 75 and 85 percent. In other words, the running loads can be predicted to run between 225 kVA and 255 kVA for such a system. Thus users should be aware that nameplate ratings indicate power for a fully featured machine and actual loadings may be significantly less for individual user's systems.

Larger systems, consisting of many components, will probably have lower demand factors when compared with smaller systems with fewer components because of inherent diversity among components.

Most computer manufacturers have preferred methods of system and equipment grounding for their hardware. Some systems require the special grounded signal reference grids with single-point grounding and strict control of the paths through which equipment ground and power system ground are interconnected. These grounds are always connected one to the other to be electrically safe and to conform with the codes. The paths may be separated up to some point. Also, since computers and office machines are accessible to people, particular attention must be given to the safety grounding of all equipment. If, for sake of radio frequency noise reduction, the equipment grounding green wire is isolated from the building/

raceways, it must be run with power conductors and carried back to the service equipment grounding conductor. Further, special semiconductive flooring may be employed for computer rooms, which enables draining off of static charges and minimizes the shock hazard to personnel. Some equipment manufacturers recommend grounding of equipment to the floor grid. In all cases, the manufacturers installation instructions should be consulted for recommendations which are specific for the installation and equipment at hand. See Section 7 for grounding of emergency and standby power systems.

Insurance companies provide loss of power and loss of production insurance which can give a cost guideline on how much can be spent to improve the power supply.

The various units of a computer system are not equally sensitive to power supply disturbances. Power requirements for equipment such as air conditioning, lights, and drive motors are much less restrictive than the logic, memory, control disk, and tape units. For economic reasons the power loads should be separated into those requiring buffer or filter action with an uninterruptible power supply and those which require no buffering and can accept a 0.5 s to 1 min power interruption while switching to a standby source of electric power.

Assuming that such a separation of computer or data processing equipment, or both, is possible, the next step would involve the selection of a system or equipment which would best solve the user's problems.

In summary, the following steps are intended to serve as a checklist toward problem solving where power problems are present. Care is required and computer manufacturer's advice should be heeded in arranging and coordinating the ground references where more than one power source supplies power to computer units in a common system.

(1) Survey power quality requirements of all installed equipment to determine maximum degree of sensitivity of the individual components. This data is often not available from manufacturer's published literature.

(2) Check to see if the local utility company has operating records. If no records are available, the user may choose to install a power disturbance monitor to detect and record transient power deficiencies at the point of usage and obtain a profile of power transients for a minimum two-week period. The results of such recordings should be used with the recognition that such seasonal conditions of lightning and other weather-related problems may not have been accounted for.

(3) Review and analyze the monitor's records to identify the magnitude, number, time, date, and characteristics of the recorded transients, interruptions, and prolonged outages.

(4) Classify the transients according to their origin: from the utility; from the operation of electrical equipment in the computer's vicinity; and from the operation of computer units and their accessories.

(5) As appropriate, prepare engineering design and cost estimates of a power-buffering system of required capacity (kVA). The chosen power-buffering system should accommodate both future expansion and possible replacement with more recently marketed computer models.

(6) Evaluate actual and intangible losses due to interruptions or malfunctions of the electronic data processing equipment from all causes, including losses attributable to power supply deficiencies or failure.

GENERAL NEED GUIDELINES

(7) Compare expected losses without power-buffering equipment to the cost of owning and operating various types of a power-buffering system as insurance against transient-caused interruptions or malfunctions of computer equipment.

(8) If the previous step justifies an investment in a power-buffering system, proceed with the project.

3.10.5 *Power Quality Improvement Techniques.* Any attempts to categorize power improvement equipment should logically attempt also to define the time duration effectiveness of the equipment. The simplest and least costly equipment is limited in effectiveness to extremely short duration power disturbances. As the time span for protection of disturbances increases to "outage" type disruptions, the sophistication and associated cost for related equipment, as a general rule, will also increase. The following broadly classifies equipment from the very short duration to an indefinite time period. A broader classification of equipment could be the distinction between static and dynamic systems.

Equipment Systems

Power conditioners
Short time (up to 15 s, but typically 0.1 s) ride-through
Extended time (up to 30 min) ride-through
Indefinite time

The time interval for most of the equipment listed can usually be extended to an indefinite time period through the use of a supplemental standby power source such as a second utility company power feeder or on site generation. To maintain power input to the load within tolerance, the supplementary equipment must be capable of "riding through" the inherent sensing and transfer time switching associated with a second incoming feeder. With standby generation systems, ride-through time must at least equal prime movers starting time, including synchronizing and transfer switching times for those systems.

Short time ride-through is accomplished with "mechanical stored energy systems," which are described in 4.6.

Extended ride-through time is achieved through operation of the equipment in conjunction with stored energy equipment, such as batteries.

In general, "power conditioning" equipment is limited to improvement of short-duration power problems. In installations where electrical "noise" is the principal problem, an isolation transformer can be an effective solution in noise attenuation.

Isolation transformers can simultaneously isolate and change voltage, for example, from 480 V three-phase to 208Y/120 V, or can have a one-to-one transformation ratio. An ordinary transformer with separate primary and secondary circuits will provide some isolation. However, effectiveness is greatly improved if they are equipped with special shielding between the primary and secondary windings. This special shielding will reduce the noise amplitude and inhibit the passage of noise through the transformer. Performance of three-phase transformers will be greatly enhanced in handling harmonic currents and unbalanced load if the transformer is connected delta primary/wye secondary on a three-legged core. Such transformers can be equipped with primary taps to adjust output voltage when input is constantly high or low. A typical tap range is +5 percent to −5 percent in 2½ percent steps.

Beyond the isolation transformer is a wide range of voltage correction or stabilizing equipment, each having dis-

Table 8
Performance of Power Conditioning Equipment

Equipment Type	Response Time	Cost (¢/VA)	Typical Internal Impedance (%)
Regulator, low impedance	11 ms	16−24	3−5
Transformer, constant voltage	25 ms	4−5	
Reactor, electronic saturable	30 ms	20	20−30
Regulator, electronic magnetic	160 ms	8−12	
Regulator, electromechanical	5+ s	5−8	3−5
Regulator, induction	5+ s	3−4	10
Tap changer, mechanical	60 s	2−3	3−5

tinct and unique operating characteristics. Commercially available equipment which would fall into the power conditioner classification are listed in Table 8. Typical reported response times are listed for each device along with the approximate uninstalled cost or range in 1978 dollars. It it obvious, because of the relatively slow response times, that the use of several of the devices would be restricted to simple voltage regulation and would have questionable value against short-term disturbances. In addition, some of the devices have inherently high internal impedance. Any sensed voltage drop is ultimately corrected by its voltage regulating capability. However, under stepped load change, the dynamic voltage change may be unacceptable for the time required for the regulator to respond. If multiple loads are to be supplied by one device and one of the loads is switched, one must be assured that the magnitude of voltage change before the regulator can react will be within acceptable limits.

Dynamic regulation and response time are the major considerations. It is stressed that because of relatively slow response times some of the equipment is effective only for voltage adjustment or correction. Also, steady-state regulation is normally published but dynamic regulation to step changes in voltage input and load current is often omitted. Without a detailed analysis of equipment operating characteristics, often from actual test data, a user could mistakenly think that he is getting protection against short-duration line input voltage disturbances and the effects of changing load. In general, equipment is available to accept line voltage variations to a 15 percent overvoltage to a 20 percent undervoltage condition with output regulated to ± 5 percent of rating. Such a regulated output is generally sufficient for electronic data processing equipment. In a situation where there appears to be a need for an ac line regulator, the application should be reviewed with the manufacturer of the data processing equipment.

The true nature of the data processing operation must be examined in combination with its extent of usage. On line systems generally require a higher degree of power protection than off line systems. Certainly a 24 h around-the-clock operation would be more vulnerable than a one-shift-a-day operation.

Whether on line or off line, the consequences of a power failure should be determined in terms of cost, value of lost data, necessity of reruns, inconvenience to and loss of revenue from customers, possible equipment damage, repair cost, and general annoyance. It is sometimes found, particularly in off line operations, that the consequences of an outage do not justify the often considerable expenditure for a protective system. The decision must be based on how much a user should pay for insurance in light of the benefits to be derived.

To evaluate the effects of outages, an estimate of the number of outages and their duration predicted over a period of at least one year should be obtained from the utility, based on their past operating experiences.

A comparison of what is predicted with what is acceptable in terms of power limits should indicate if supplemental protection is needed.

Assuming that the predicted available power is not acceptable in terms of duration or frequency of disturbances or disruption, it must then be determined how much protection is appropriate and what solutions are available to the user.

Table 9 has been prepared to indicate what effect three types of power line disturbances will have on data processing and computer equipments. Also shown for each type of disturbance are several of the available solutions which can be applied for enhancement of the power quality. Note that in this comparison the disturbance durations are for less than 1 cycle, from 1 cycle to 10 s, and for more than 10 s. Table 10 presents similar data in a slightly different manner and shows, for each of the categories of disturbances, how effective each of the available solutions will be as a power quality improvement measure.

An indication of reliability can be expressed in hours of mean time between failure (MTBF). Data for calculating the MTBF can be obtained from Mil Standards Handbook or from ANSI/IEEE 500–1977. Equipment and system reliability can be enhanced by redundant systems and components and provision for isolation or bypass of defective parts or subassemblies.

A typical reliability prediction assumes that if the uninterruptible power supply system fails, there is to be a high probability that the static switch will operate and that main power of suitable quality will be available to handle the load. The probability of simultaneous failures of the uninterruptible power supply and static switch or main power yields an exceptionally low failure rate corresponding to MTBFs of 10 years or longer.

As with reliability, data on equipment life expectancy is difficult to obtain. Often, because of the rapid advancement of data processing technology, the protected equipment obsolescence will generally occur before the life of the protecting equipment is exceeded.

Dynamic or rotating apparatus is capable of withstanding short time and sustained overloads better than static equipment. Further, most static assemblies are more vulnerable to shock or high-inrush loads than dynamic equipment. Dynamic or rotating equipment does not lend itself well to expansion through paralleling due to complex control problems. Conversely, most static systems are easily expanded through the addition of parallel modules or additional batteries.

Because some systems have inherent lower overall efficiencies, increased heat losses will be evident. The first cost of supplemental cooling and ventilating equipment, with ongoing operating and maintenance costs, must be considered. Further, the need of additional auxiliary or standby power capacity to drive the cooling and ventilating equipment, when normal power curtailment occurs, must also be examined.

Most equipment will occupy building

Table 9
Summary of Typical Power-Line Disturbances

Type of Voltage Disturbance	Voltage Level of Disturbance	Duration of Disturbance	Typical Effects on Computer Equipment	Typical Power Enhancement Projects
Outage	Below 85% VRMS	More than 10 s	Built-in voltage sensors will power down computer equipment in an uncontrolled manner. Processing is interrupted usually resulting in excessive restart/rerun time, possible loss of data, or damage to hardware.	Uninterruptible power supply system, standby diesel generators, dual power feeders, general improvements to power distribution system.
Momentary under- and overvoltages (sags and surges)	Below 85% VRMS and above 105% VRMS	From 16.7 ms (1 cycle) to 10 s	Equipment may power down depending on duration and magnitude of disturbance. If so, processing is interrupted usually resulting in excessive restart/rerun time. In severe cases loss of data and damage to hardware may occur.	Solid-state switching between dual feeders, motor generator sets, fast response line voltage regulators, balance computer load on three-phase power, improve computer equipment grounding, general improvement to power distribution system.
Transient overvoltages (impulses or spikes)	100% VRMS or higher (measured as instantaneous voltage above or below the line VRMS)	Less than 16.7 ms (1 cycle)	Data disruptions leading to errors, unready indications, etc, may cause individual equipment to stop processing. However, direct effects on the system are not normally detectable. Rarely, a severe transient will cause equipment to power down. Damage to electronic components may also occur if the equipment is not properly grounded or otherwise protected from transient overvoltages.	Isolation transformers, transient suppressors, power-line filters, primary and secondary lightning arrestors, balance computer load, improve computer equipment grounding.

GENERAL NEED GUIDELINES

Table 10
Relative Effectiveness of Power Enhancement Projects in Eliminating or Moderating Power Disturbances
(US Navy)

Disturbance Type	Uninterruptible Power Supply (UPS) System and Standby Diesel Generator	Uninterruptible Power Supply (UPS) System	Dual Power Feeders		Motor Generator	Solid-State Line Voltage Regulator	Specialty Shielded Isolating Transformer	Suppressors, Filters, and Lightning Arresters	Balance Computer Load on 3-Phase Supply, Improve Grounding
			Secondary Spot Network	Secondary Selective*					
Transient and oscillatory overvoltage	All source-caused transients and no load-caused transients	All source transients and no load transients	None	None	All source transients and no load transients	Most source transients and no load transients	Most source transients and no load transients	Most	Some**
Momentary undervoltage or overvoltage	All	All	None	Most	Most	Some (depends on response time)	None	None	Some**
Outage	All	Only outages of a duration equal to the discharge time of the battery	Most	Most	Only "brownout"	Only "brownout"	None	None	None

*Includes special application of a solid-state static switch between two independent sources.
**These improvements do not eliminate or moderate power-line disturbances, but they do make the computer equipment significantly less susceptible to under- and overvoltages. Assistance of the computer manufacturer is generally required to identify grounding problems.

interior space, which not only costs money but will require environmental treatment or tempering. Such space is often treated as leasable space and as such its cost considered as an ongoing cost in any comparative analysis.

Delivery, setup, checkout, and debug time can often be an important and deciding factor when the dependence and prime emphasis is placed on having a data processing system or computer and its capabilities in service on an around-the-clock basis.

Some of the solutions for improvement of power quality will require the installation of an auxiliary power source with some form of prime mover. Other sections in this text provide detailed guidance toward a prime mover selection. The most popular choice is the diesel engine, with gas turbine usage increasing. Often it is to the advantage of the user to provide for a separation of the source for the data processing equipment or its power improvement equipment from the source for the facility's environmental equipment. Since the only objective of the auxiliary power source is to provide an almost immediate source of power, reliability in starting is the most important characteristic. Usually the unit is called on to start after a sustained period of idleness. The time period for starting could be important since the equipment could be called on to operate in conjunction with a stored energy type of ride-through system. Often data processing equipment is installed in a relatively quiet atmosphere so that whichever choice of prime mover is made, some form of acoustic treatment will probably be required.

3.11 Life Safety and Life Support Systems

3.11.1 *Introduction.* Emergency power for life safety systems is required in many different types of commercial and industrial facilities. For example, an alternative power source is needed to illuminate exits and to operate alarm systems in most public buildings. These needs generally relate to fire safety as set forth in NFPA No 101-1976, Life Safety Code.

In contrast, emergency power requirements for life support systems, such as a heart-lung machine or medical diagnostic equipment, are generally limited to health-care facilities. These requirements result from operational needs unique to the health care environment.

Emergency power for both life safety and life support equipment is addressed in this section. In doing so, the emphasis is intended to be primarily in the viewpoint of the user.

3.11.2 *Health-Care Facilities.* Typical examples of both life safety and life support emergency power requirements are found in large health-care facilities. An appreciation for both of these needs can therefore be obtained by reviewing a hospital's emergency power system.

Hospitals are becoming increasingly dependent on electrical apparatus for patient life support and treatment as illustrated by the following:

Electronic monitoring systems are being used to conduct therapy for critically ill patients. In the operating suite, during open-heart surgery, an electric heart-lung machine maintains extra corporeal circulation. In the intensive care unit patients depend on ventilators powered by electricity. Cardiac-assist devices augment a patient's own circulation in the coronary care unit. Continuous lighting is needed to observe patients in primary care areas and power is needed to maintain refrigerated storage of vital supplies such as blood and tissue banks.

In addition to these examples of direct dependency on electric power in patient care, the sustained loss of electrical power can also result in an undesirable traumatic experience for some seriously ill patients.

Hospitals also must have a highly reliable supply of emergency power for life safety systems to ensure that the lives of sick or disabled persons are protected during emergencies.

Interruption of normal electrical services to hospitals may be caused by a variety of natural phenomena. These include storms, floods, fires, earthquakes, explosions, traffic accidents, electrical equipment failure, and human error. Several large area blackouts have been experienced in the recent past with prolonged local outages. As a consequence of the "energy crisis" the incidence of such failures is thought by many authorities to be on the rise. Alternative sources of power for supplying vital life safety and life support systems must be provided to protect patients relying on these systems.

The acceptable duration for an interruption of normal power service to critical hospital loads is the subject of many state codes and regulations (see Table 1). National standards often referenced by the states and specifically addressing this issue are NFPA No 76A-1977, Essential Electrical Systems for Health-Care Facilities, and Article 517 of NFPA No 70-1978, National Electrical Code.

NFPA No 76A-1977 requires that all health-care facilities maintain an alternate source of electrical power. With few exceptions, this source must be an on-site generator capable of servicing both essential major electrical equipment and emergency systems.

For hospitals, NFPA No 76A-1977 provides the following criteria with respect to the emergency system: "Those functions of patient care, depending on lighting or appliances that are permitted to be connected to the Emergency System are divided into two mandatory branches; the Life Safety and the Critical. The branches of the Emergency System shall be installed and connected to the alternate power source . . . so that all functions specified . . . shall be automatically restored to operation within 10 seconds after interruption of the normal power."

To meet the "10 second criteria" the emergency system must include independent distribution circuits with automatic transfer to the alternate power source. Two-way bypass and isolation transfer switches are recommended for the emergency branches. Fig 5 shows the emergency system wiring arrangement from a typical hospital. The hospital emergency system installation should follow Article 700 of NFPA No 70-1978.

The "life safety branch" of the emergency system, as described in NFPA No 76A-1977, includes illumination for means of egress and exit signs (NFPA No 101-1976 requirement), fire alarms and systems, alarms for nonflammable medical gas systems (NFPA No 56F-1977 requirement), hospital communication systems, and task illumination/selected receptacles at the emergency generator set location.

NFPA No 76A-1977 contains a complete listing of circuits to be connected to the critical branch feed areas and functions related to patient care. For most of these critical loads the "10 second criteria" is considered to be sufficient. However, an instantaneous restoration of minimal task lighting, using battery systems, is recommended in operating, delivery, and radiology rooms where the loss of lighting due to power failure might cause severe and immediate danger to a patient undergoing surgery or an invasive radiographic procedure.

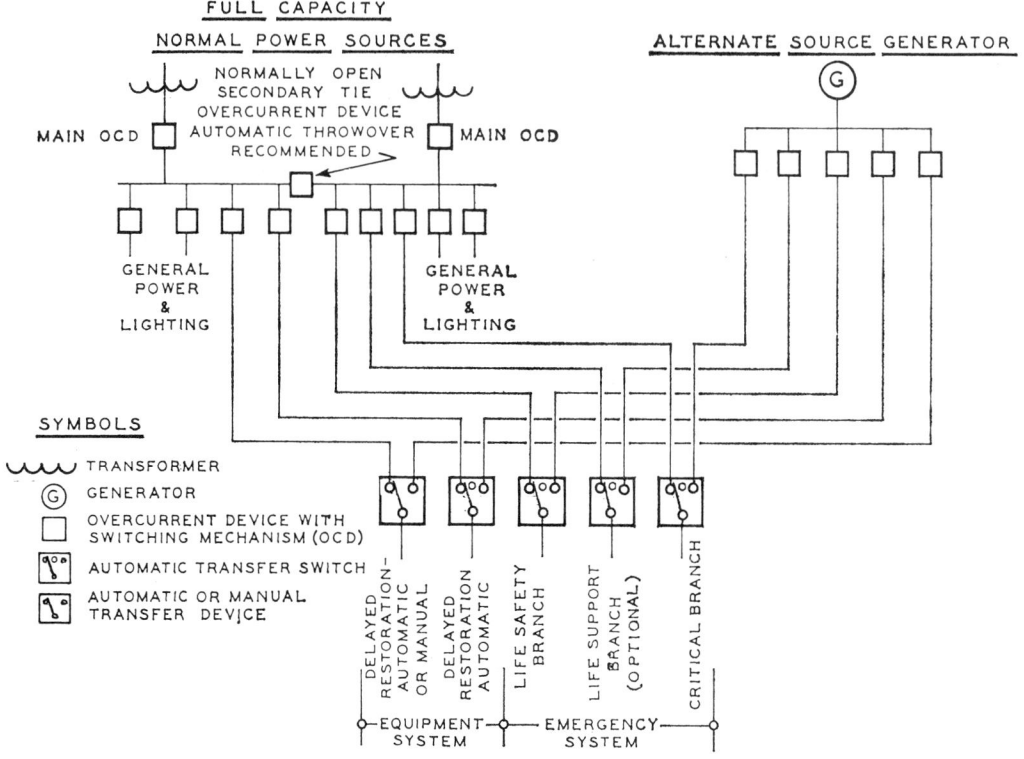

Fig 5
Typical Hospital Wiring Arrangement
NFPA No 76A-1977

Examples of the types of life support and life safety equipment available with built-in battery backup power are the following:

Life Support	Life Safety
Aortic balloon pumps	Fire monitoring and alarm system
ECG and EEG monitors	Communication systems
Portable defibrilators	Safety lighting
Portable respirators	
Task lighting	

Periodic tests and maintenance are essential to assure the reliability of the alternate source of power and other elements of the emergency system. Regular performance of maintenance is subject to the approval of the authority having jurisdiction. A record of procedures must be kept. Power outages in the past have shown the uselessness of having unmaintained emergency and standby systems which fail in time of need. Basic maintenance and test requirements for essential electrical systems in hospitals are also included in NFPA No 76A-1977.

Since emergency and standby power systems are mandatory in most areas, the hospital electrical system designer must review the state and city building laws and regulations. These laws often reference NFPA No 70, 76A, and 101 as well as other national standards such as the Uniform Building Code. Insurance and building inspectors should also be consulted. In addition, installation, operation, and maintenance of the emergency power system is a factor in accreditation by the Joint Commission on Accreditation of Hospitals.

There is another power problem in hospitals that is not covered by any of the previously referenced codes and standards. It is the adverse effect of poor power quality on sensitive electronic loads. With the growing dependence on electronic equipment in medical facilities users are finding that the "10 second criteria" no longer provides an acceptable functional capability. These loads are disturbed by transient overvoltages that are measured in microseconds and dips in steady-state voltage lasting only a few cycles of the 60 Hz wave. See 3.10 for a detailed description of typical power line disturbances and their effect on sensitive electronic equipment.

Some examples of sensitive hospital loads and the probable results of poor power quality (evidenced by powerline disturbances) are shown in Table 11.

This problem can be diminished if the manufacturer of sensitive or critical medical equipment thoughtfully designs the equipment to tolerate power-line disturbances. For example, built-in batteries can power equipment or at least maintain critical data during momentary interruptions in the normal supply. In addition, the susceptibility to transient overvoltage can be reduced if the equipment power supply design includes proper isolation and built-in transient overvoltage protection. These preventive measures by the medical equipment manufacturer are considered to be an efficient method of avoiding power quality-related problems.

Computer systems adapted to hospital applications typically do not have built-in protection from power-line disturbances. Nevertheless, just as in the business sector, computer systems will inevitably play a dominant role in hospitals. Organizations such as the Society for Computer

Table 11
Sensitive Hospital Loads

Sensitive Load	Probable Result of Severe Power Disturbances
ICU patient monitoring systems	Incorrect trend analysis or false alarm Lost time for restart and reprogramming Loss of data
Lab equipment blood gas analyzers	Depending on complexity, may result in extensive reprogramming and set up lost time
Blood cell counter	Automatic power down for self-protection which disrupts or delays test runs
Nuclear monitoring	Disruption of test and inability to retest due to patients' radiation exposure limit
X-ray/ultrasonic scanner	Varies from no effect to breakdown depending on equipment susceptibility
Hospital information systems (HIS) computers	Typical computer subject to disruption resulting in restart and reprogramming lost time and loss of memory (see 3.10)

Medicine provide a forum for identifying and putting into practice hospital computer applications. These applications have been summarized as "patient care computing" and include patient monitoring, laboratory, radiology, diagnostic support, patient data base acquisition, patient scheduling, and so on [1].

With the increased dependence on computer systems in hospitals, more priority will certainly be placed on supplying continuous and disturbance-free power in the future. For computers and other life safety and life support systems where it is impractical for the equipment manufacturer to provide built-in solutions to power-quality problems, the user may find the need for power interface systems in his facility. Power interface systems bridge the gap between equipment susceptibilities and powerline disturbances. For overvoltages or electrical noise problems, isolation transformers or transient suppressors may be an effective deterrent when installed in the building power source. Required stored energy to ride-through low voltage and momentary interruptions may be obtained from a power buffering motor-generator set or, when the need for power continuity warrants, an uninterruptible power supply (battery system). These interface systems are addressed in Section 4.

A large health-care facility involved in extensive life support related functions probably will require a combination of both emergency power and power interface systems. In this case, a central system approach to satisfy special power requirements may be more cost effective than one that addresses individual equipment separately.

It is strongly recommended that the user with an apparent power-quality problem consult and secure the services of qualified engineers familiar with the unique power-quality requirements of his facility.

3.11.3 *Other Critical Life Systems.* Other critical life systems not necessarily unique to hospitals are subject to similar special laws and regulations regarding emergency power. Some examples follow:

(1) Controls for pressure vessels, such as boilers or ovens, where a failure may result in a life endangering explosion or fire

(2) Air supply systems for persons in a closed area

(3) Fire pumps, alarms, and systems

(4) Communications systems in hazardous areas

(5) Industrial process in which the interruption of power would create a hazard to life

To meet these or other specialized emergency power requirements, the assistance of an engineer who is experienced in the particular problem area should be obtained.

3.12 Communication Systems

3.12.1 *Description.* Communication systems are those facilities which require electric power for verbal, written, or facsimile transmission and reception. Common systems of this type are:

(1) Telephone
(2) Teletypewriter
(3) Paging
(4) Radio
(5) Television

Needs of one or all of the above communication systems during a power failure may well justify the cost of one or more emergency and standby power systems, possibly in conjunction with any other critical loads.

3.12.2 *Commonly Used Auxiliary Power Systems.* Battery or battery and converter equipment are practical sources of power using a float-charging system.

Small engine generators are practical and economical. In the size ranges usually required from 1 to 5 kW, the installed cost ranges from about $150 to $300 per kilowatt, depending upon the quality of the equipment, battery life, gasoline storage problems, and amount of automatic equipment deemed to be necessary or convenient.

3.12.3 *Evaluating Need for Auxiliary Power System.* The need of an emergency or standby power system for communications should include satisfactory answers to the following questions:

(1) Will the communication equipment be required

   (a) To issue orders for an orderly shutdown of processes and equipment?

   (b) To announce instructions to personnel? A typical announcement might be to wait by the machines or to check out for the duration of the shift.

   (c) To call for help, issue warnings, and coordinate the work should there be a fire, civil disturbance, vandalism, or other threat to personnel safety or plant security?

(2) How will vital messages be received or sent for remote plants concerning production, inventory, or sales changes?

(3) How will key personnel be found, or instructed; and how will these persons report conditions to a central responsible source of control?

Many other questions may be asked, but the maintaining of communications under emergency conditions will save vital time and expedite the return to normal conditions with reduced confusion.

For industrial plants, the telephone system is usually powered both regularly and on a standby and emergency basis by the telephone company, normally by batteries or by standby generation. In some plants there is a separate in-plant telephone system which may be powered by batteries under a float charge which will maintain the communication system for several hours during a prolonged power interruption. The user should check to see that his system will function for the required time with loss of normal power.

Usually various bells, horns, and other call devices connected to the telephone system are powered by the lighting circuits and will stop functioning if there is a power interruption, even though the receiver and transmitter work. If required, these should be wired to the backup power system.

Teletype equipment functions much the same as the telephone system so that the signal may be present but the printout device will not operate without local power.

Paging systems in plants are frequently extensive, using many hundreds of watts of audio power and several kilowatts of 120 V power. The need to use the system during a power interruption should not be overlooked.

Radio systems are common in industrial plants. While the mobile units for personnel as well as in-plant and out-of-plant car and truck units are generally self-powered by batteries, the main base station usually is connected to the nearest commercial power source. While mobile units may or may not be able to speak to each other, the system as a whole usually stops functioning if the base station power fails. Consideration of emergency and standby power for this base station should not be overlooked.

In some plants and commercial buildings, paging or broadcasting is done by dialing through the telephone system. In this case, both the telephone and paging or radio systems should be supplied with emergency power.

3.13 Signal Circuits

3.13.1 *Description.* A signal circuit supplies power to a device which gives a recognizable signal. Such devices include bells, buzzers, code-calling equipment, lights, horns, sirens, and many other devices.

3.13.2 *Signal Circuits in Health-Care Facilities.* Signal circuits in medical buildings which should be provided with continuous emergency power within 10 s [NFPA No 101-1976] include

(1) Fire alarm systems:
   Manually initiated
   Automatic fire detection
   Water flow alarm devices used with sprinkler systems
(2) Alarms required for systems used for piping of nonflammable medical gas
(3) Paging system
(4) Nurse's station signaling system from patient areas
(5) Alarm systems attached to equipment required to operate for the safety of major apparatus
(6) Signaling equipment for elevators in buildings of more than four stories

3.13.3 *Signal Circuits in Industrial and Commercial Buildings.* Signal circuits for commercial buildings and industrial plants which may require continuous emergency power within 1 min include

(1) Fire alarm system
(2) Watchman's tour system
(3) Elevator signal system
(4) Door signals (into restricted areas such as boiler rooms and laboratories with electric door locks)
(5) Liquid level, pressure, and temperature indications

3.13.4 *Types of Auxiliary Power Systems.* The emergency supply for the signal circuits can be (1) engine-driven generators, (2) multiple utility services, or (3) floating battery systems with auxiliary power. Most signal circuits operate down to 70 percent rated voltage and therefore require no special voltage-sensing relays on the transfer device.

It is recommended that an emergency source of electric power be supplied for every part of a fire alarm and security system. A local battery supply on float charge, close to the power need, in continuous service is very reliable. A usually acceptable substitute is an automatic transfer to a battery system when prime power fails.

Signal circuits are usually a small electrical burden and integral part of a total load that also requires an emergency source. Therefore, the selection of emergency system and hardware generally depends upon the requirement of other related loads.

3.14 Standards References

The following standards publications were used as references in preparing this section.

ANSI C84.1-1977, Voltage Ratings for Electric Power Systems and Equipment (60 Hz)[1]

ANSI/IEEE 500-1977, Guide to the Collection and Presentation of Electrical, Electronic, and Sensing Component Reliability Data for Nuclear Power Generating Stations[2]

EGSMA GDT2-1971, Glossary of Standard Industry Terminology and Definitions[3]

ANSI/IEEE Std 100-1977, Dictionary of Electrical and Electronics Terms

---

[1]American National Standards Institute, 1430 Broadway, New York, NY 10018.
[2]Institute of Electrical and Electronics Engineers, Inc, 345 East 47 Street, New York, NY 10017.
[3]Electrical Generating Systems Marketing Association, Tribune Tower, 435 N Michigan, Chicago, IL 60611.

IEEE Std 141-1976, Electric Power Distribution for Industrial Plants

IEEE Std 241-1974, Electric Power Systems in Commercial Buildings

NFPA No 30-1973, Flammable Combustible Liquids Code[4]

NFPA No 56F-1977, Non-Flammable Medical Gas Systems

NFPA No 70-1978, National Electrical Code

NFPA No 76A-1977, Essential Electrical Systems for Health Care Facilities

NFPA No 101-1976, Life Safety Code

Public Law No 91-596, Williams-Steiger Occupational Safety and Health Act of 1970 (84 Stat 1593, 1600; 29 USC 656, 657), Chapter XVII of Title 29 of The Code of Federal Regulations, established on April 13, 1970 (36 FR 7006) as amended by adding thereto a new part 1910. Washington, DC: Department of Labor.

Amendments to Williams-Steiger Occupational Safety and Health Act of 1970, as published in the Federal Register. Washington, D.C.: Office of the Federal Register, National Archives and Records Service, General Services Administration.

Title 24, California Administrative Code, Part 3, Basic Electrical Regulations, Article E700, Emergency Systems. Sacramento, CA: Document Section.

3.15 References and Bibliography
3.15.1 References

[1] *Readers Digest Almanac.* Pleasantville, NY: The Readers Digest Association, 1972, p 790.

---
[4]National Fire Protection Association, 470 Atlantic Avenue, Boston, MA 02210.

[2] CASTENSCHIOLD, R. Criteria for Rating and Application of Automatic Transfer Switches. *IEEE Conference Record of 1970 Industrial and Commercial Power Systems and Electric Space Heating and Air-Conditioning Joint Technical Conference,* IEEE 70C8-IGA, pp 11–16.

[3] KAUFMAN, J. E., Ed. *IES Lighting Handbook.* New York: Illuminating Engineering Society, 1972, sec 14, pp 14–10 to 14–12 (Emergency Lighting).

[4] FINK, D.G., and CARROLL, J. M. *Standard Handbook for Electrical Engineers,* 10th ed. New York: McGraw-Hill, 1968, pp 20–8, 20–9.

[5] GILBERT, M. M. What the Chemical Industry Can Do to Minimize Effects from Electrical Disturbances. Paper CP 60-1161, presented at the AIEE and ASME National Power Conference, Philadelphia, PA, Sept 21–23, 1960.

[6] IEEE Committee Report. *Protection Fundamentals for Low-Voltage Electrical Distribution Systems in Commercial Buildings.* IEEE JH 2112-1, 1974, pp 19, 20.

3.15.2 Bibliography

[7] KATZ, E. G. Evaluation of Hospital Essential Electrical Systems. *Fire Journal,* vol 62, Nov 1968.

[8] BEEMAN, D. L., Ed. *Industrial Power Systems Handbook.* New York: McGraw-Hill, 1955.

[9] BURCH, B. F., JR. Protection of Computers against Transients, Interruptions, and Outages. Presented at the 1967 IEEE Industry and General Applications Group Annual Meeting, Pittsburgh, PA, Oct 4, 1967.

[10] FERENCY, N. Is a UPS Really Necessary? *Electronic Products Magazine*, Apr 17, 1972, pp 126–127.

[11] FISCHER, E. I. Emergency Power Facilities for a Research Laboratory. *Conference Record of the 1966 IEEE Industry and General Applications Group Annual Meeting*, IEEE 34-C36, pp 305–324.

[12] HEISING, C. R., and DUNKI-JACOBS, J. R. Application of Reliability Concepts to Industrial Power Systems. *Conference Record of the 1972 IEEE Industrial Applications Society Annual Meeting.* IEEE 72CHO685-81A, pp 289–295.

[13] KNIGHT, R. L., and YUEN, M. H. The Uninterruptible Power Evolution—Are Our Problems Solved? *Conference Record of the 1973 IEEE Industry Applications Society Annual Meeting*, IEEE 73CHO763-31A, pp 481–485.

[14] MILLER, N. A. Noninterruptible Power Supplies for Essential Control Systems in Power Plants. Presented at the IEEE IAS–IEC Group Meeting on Heating, Chicago Chapter, Jun 7, 1972.

[15] OLIVER, R. L. Plant Engineering at RCA. *Plant Engineering*, Apr 1969.

[16] Beating the Blackouts. *The Wall Street Journal*, Jul 21, 1970.

[17] The On-Site Power Market. *Electrical Construction and Maintenance*, Jan 1971, pp 57–71.

[18] JENKIN, M. A. Functions of Patient Care Computing. *Medical Instrumentation*, vol 12, July–Aug 1978.

[19] KEY, T. S. Diagnosing Power Quality Related Computer Problems. *IEEE I–CPS Tech Conf Record*, 78CH1302-91A, Jun 1978, pp 48–56.

[20] TUCKER, R. Line Voltage Regulators, Less Costly Alternative to UPS. *Electrical Consultant*, Aug 1974.

4. Systems and Hardware

4.1 Guidelines for Use

When a user experiences an equipment problem due to failure of the electric power supply, he must either live with the problem, change his equipment or system to perform satisfactorily during a failure of the existing power supply, or alter the supply to prevent potential failures. In many cases, the correct decision is to change the equipment or system, but this publication does not examine these cases.

Once the electric power user's study has shown that the correct approach is to alter or supplement the power supply source, a study should be undertaken to determine the proper systems and hardware which will meet the need at the lowest cost for the electric power required.

This publication describes combinations of systems and hardware which will overcome the following types of electric power failures with good reliability:

(1) Long-time interruption (hours)
(2) Medium-time interruption (minutes)
(3) Short-time interruption (seconds)
(4) Transient interruption (milliseconds)
(5) Over- or undervoltage
(6) Over- or underfrequency
(7) Transients in the prime source of power
(8) Transients caused by the user's equipment

Emergency power systems are of two basic types: (1) an electric power source separate from the prime source of power, operating in parallel, which maintains power to the critical loads should the prime source fail; or (2) an available reliable power source to which critical loads are rapidly switched automatically when the prime source of power fails.

Emergency systems are frequently, but not always, characterized by continuous or rapid availability of electric power of limited-time duration and supplied by a separate wiring system. Frequently the emergency power system has a standby power system available which increases the emergency supply time to as long as needed.

Standby power systems are made up of the following main components:

(1) An alternate reliable source of electric energy separate from the prime power source
(2) Starting and regulating control if on-site standby generation is selected as the source

(3) Controls which transfer loads from the prime or emergency power source to the standby source

It is prudent for the user to establish his practical need from the previous sections of this publication before specifying and purchasing, since costs will be found to rise as the following demands are specified for systems and hardware:

(1) Longer equipment life
(2) Increased capacity
(3) Closer frequency regulation
(4) Closer voltage regulation
(5) Freedom from voltage or frequency transients
(6) Increased availability
(7) Continuous operation of an uninterruptible power supply system
(8) Increased reliability
(9) Increased temporary overload capability
(10) Quiet operation
(11) Safety from fuel hazards
(12) Pollution-free operation
(13) Freedom from harmonics
(14) Close voltage and frequency regulation with wider range rapid load changes

An allowance should be made for load growth. Future power requirements frequently need to be connected to the emergency and standby system. It may be desirable to add additional existing power loads to the more reliable power bus as soon as the advantages are realized in practice. If additional capacity cannot be justified initially, the equipment and system should be selected and designed for future economic expansion compatible with the initial installation.

Operating costs of the systems and hardware are usually secondary to meeting the need, but should be included as a factor in the selection. These include cost of fuel, inspection frequency, ease of maintenance, frequency of testing, cost of parts, and taxes.

Installation quality should be high to prevent losing the reliability of electric power designed into the system and purchased in the hardware [1].[1] One must guard against introducing voltage transients into the emergency and standby power distribution system. Satisfactory voltage levels must be maintained under all loading conditions.

For industrial plants, electrical systems should conform to IEEE Std 141-1976. For commercial buildings, electrical systems should conform to IEEE Std 241-1974. Grounding practices should follow the recommendations in ANSI/IEEE 142-1972. Additional shielding, bonding, grounding, and even filtering may be required to maintain the quality of the emergency and standby power supply.

When energized sources of power are held available for emergency use, there should be included in the system a light to show that the source is energized and an alarm to signal loss of available power. These should be located where responsible persons can take action, should an alarm sound. An alternate utility source (4.3) or battery system (4.8) are typical on line sources normally requiring these signals.

Users should take these additional steps to assure performance reliability:

(1) Establish regular inspections using a check sheet and recording exceptions
(2) Perform regular preventive maintenance and repair items of exceptions found during the inspections
(3) Set up a trial at regular intervals simulating a power failure but timed so as not to encounter hazards or losses should the system not operate as anticipated

There are mathematical methods for quantitatively determining the reliability of an emergency or standby power system.

---

[1]Numbers in brackets correspond to those in the References at the end of this Section.

Once the need is established, it may be advisable to calculate the system reliability, especially for emergency systems involving possible injury or loss of life. One such method is detailed in Sawer [2] and Heising and Johnston [3].

In 4.2 through 4.8 the systems and hardware in practical use today are presented. On the horizon are other sources of electric energy and systems which may soon serve to fill some of the needs for emergency and standby power systems. Some of these sources and systems foreseen for future applications are:

(1) Fuel cells which convert chemical energy directly into electric energy

(2) Radiant and solar cells which convert radiant energy into electric energy

(3) Chemical luminescence arranged to convert energy into light

(4) Nuclear power generators

(5) Thermocouples which convert heat to electric energy

(6) Radio isotopes which excite chemical panels to produce illumination

Recommended practices for the application of these and other developments will be presented as they become developed to the state of practical application to fill the user's needs as detailed in the previous sections.

4.2 Engine-Driven Generators

4.2.1 *Introduction.* These units are "work horses" which fulfill the need for emergency and standby power. They are available from small 1 kVA units to those of several thousand kVA. When properly maintained and kept warm, they dependably come on line within 8 to 15 s. In addition to providing emergency power, engine-driven generators are also used for handling peak loads and are sometimes used as the preferred source of power.

They fill the need of back-up power for uninterruptible power systems. Where well regulated systems, free from voltage, frequency, or harmonic disturbances, are required, such as for computer operations, a "buffer" is usually needed between the critical load and the engine-driven generators.

4.2.2 *Diesel-Engine Generators.* A typical diesel-engine-driven generator rated 500 kW is shown in Fig 6. Typical ratings are given in Table 12. Installed operating units with fuel supplies (other than small portable units) will cost between $100 to $270 per kW. Lower speed units are heavier and more costly, but are more suitable for continuous duty.

Diesel engines are somewhat more costly and heavier in smaller sizes, but are rugged and dependable. The cost of fuel is lower and the fire and explosion hazard is considerably lower than for gasoline engines. Sizes vary from about 2.5 kW to 1100 kW.

4.2.3 *Gasoline-Engine Generators.* Gasoline engines are satisfactory for installations up to about 100 kW output. They start rapidly and are low in initial cost as compared to diesel engines. Disadvantages are a higher operating cost, a greater hazard due to the storing and handling of gasoline, short storage life of the fuel, and generally a lower mean-time between overhaul.

4.2.4 *Gas-Engine Generators.* Natural gas and LP gas engines rank with gasoline engines in cost and are available up to about 600 kW. They provide quick starting after long shutdown periods because of the fresh fuel supply. Engine life is longer with reduced maintenance because of the clean burning of natural gas. However, consideration must be given to the possibility of both the electric utility and the natural gas supply being unavailable concurrently. To compare directly with gasoline, the natural gas must have a heat value of at least 1100 Btu/ft^3. Considerations in selecting natural or LP

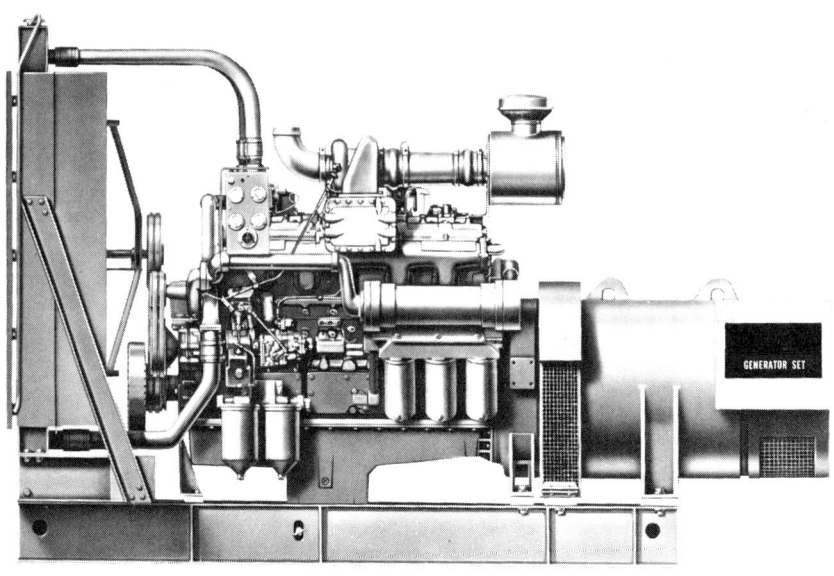

**Fig 6
Typical Engine-Driven Generator; Diesel,
Gasoline, or Gas Fueled**

gas fueled engines are the availability and dependability of the fuel supply, especially in an emergency situation.

4.2.5 *Derating Requirements.* As noted in 4.4, altitude will cause a serious derating of the prime mover to deliver the torque required for the full generator output unless a supercharger has been added. Oversizing of the engine is a must for higher altitudes. The generators are less critical in output capacity if adequate cooling air is supplied to carry away the heat generated by the losses.

A general rule for derating engine power loss with altitude increase is to derate about 4 percent for each 1000 ft increase in altitude. Also, an average derating factor for high ambient temperature is 1 percent for each 10°F above 60°F. Temperature derating is not considered as important as altitude derating.

4.2.6 *Multiple-Engine Generator Set Systems.* Automatic starting of multiple units and automatic synchronizing controls are available and practical for multiple-unit installations. Advantages of several smaller units over one large unit should be considered since emergency and standby power can be available while one unit is being maintained or overhauled. While starting is usually reliable, if the units are warm and maintained by regular exercising, the likelihood of all the units not starting is extremely low as compared to a single unit.

Smaller units also allow the "building block" concept. As capital is made available and the need of increased capacity grows, additional units of identical size and type may be added, thus simplifying the parts, maintenance, and training problems. The trend toward larger

Table 12
Typical Ratings of Engine-Driven Generators
Approximate 1978 Prices

Nominal Rating (kW)	Continuous Duty Rating (kW)	Standby Rating (kW)	Power Factor	Prime Mover			Speed (r/min)	Cost ($)
				Gasoline	Diesel	Natural Gas/ LP Gas		
5	5	5	1.0	×		×	1800	1 900
10	10	12.5		×			1800	3 200
25	25	30	0.8		×		1800	6 800
100	90	100	0.8		×	×	1800	13 000
250	200	250	0.8		×	×	1800	23 000
750	665	730	0.8		×		1200	82 000
1000	875	900	0.8		×		1200	110 000
1000	975(G) 800(D)	1100	0.8		×	×	1200	130 000 120 000

emergency power supplies also justifies the use of multiple sets to provide the additional power. Figs 7–10 illustrate typical multiple-set systems.

One very important consideration in the selection of "smaller" versus "larger" units is the service to which they will be subjected. Smaller units are nearly all higher speed (1800 r/min) engine-driven sets; and if these units are to be run continuously for long periods of time, the engine should be evaluated very thoroughly. Experience has shown that many 1800 r/min engine-driven sets are not adequate as continuous power supplies (several weeks' duration). Everything else being equal, the engine selected for a continuous duty application should be the one with the largest piston displacement, the lowest piston speed, and the lowest brake mean effective pressure or the average pressure on the piston over one complete engine cycle at rated output. If a small continuous-duty engine generator set is required, consideration might be given to a special lower speed unit.

4.2.7 Construction and Controls. The basic electrical components are the engine generator set and associated meters, controls, and switchgear. Most installations include a single generator set designed to serve either all the normal electrical needs of a building or a limited emergency circuit. Sometimes the system includes two or more generators of different types and sizes, serving different types of loads. Also, two or more generators may be operating in parallel to serve the same load.

4.2.8 Typical Engine Generator Systems. This information is designed to help in selecting the electrical components of a generator installation. No attempt has been made to cover every situation that might arise.

In Figs 7–11 the following abbreviations are used:

ATD = Automatic transfer device (automatic transfer switch or electrically operated circuit breaker)
CB = Circuit breaker
EG = Engine-driven generator set
LDC = Load-dumping contactor, electrically operated, mechanically held

Fig 7 shows a standby power system where if power fails from the normal

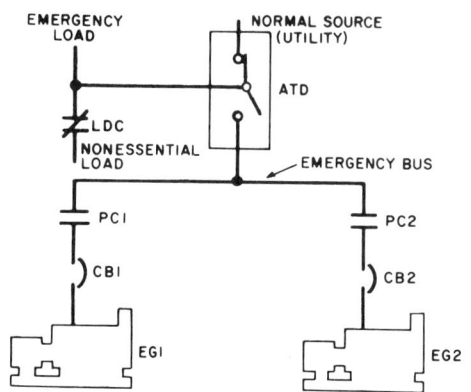

Fig 7
Two Engine Generator Sets Operating in Parallel

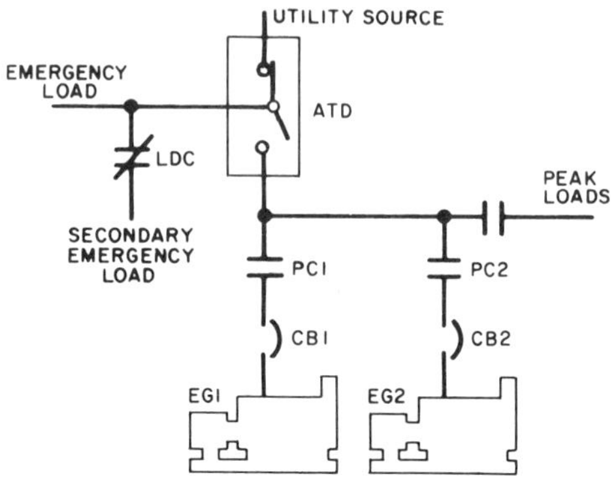

Fig 8
Peaking Power Control System

source, both engines automatically start. The first generator to reach operating voltage and frequency will actuate load dumping circuits and cause the remaining load to transfer to this generator. When the second generator is in synchronism, it will be paralleled automatically with the first. After the generators are paralleled, all or part of the dumped load is reconnected if the standby capacity is adequate.

If one generator fails, it is immediately disconnected. A proportionate share of the load is dumped to reduce the load to where the remaining generator can handle it. When the failed generator is reinstated, the dumped load is reconnected.

When the normal source is restored, the load is retransferred and the generators are automatically disconnected and shut down.

With the system shown in Fig 8, idle standby generator sets can perform a secondary function by helping to supply power for peak loads. Depending on the load requirements, this system starts one unit or more to supply peak loads while the utility service supplies the emergency loads. When the second generator is in synchronism, it will be paralleled automatically with the first. If the utility service fails, the peak loads are automatically disconnected and the generators pick up the emergency loads through the transfer device.

Fig 9 shows a standby power system where there is a split emergency load with one load being more critical than the other.

When the prime source fails, both generators start. If load 1 is the preferential load, the generator that reaches operating speed first is put on the line by automatic transfer device 3 to feed load 1 through automatic transfer device 1. When the other generator reaches operating speed, it then feeds load 2. If the generator feeding load 1 fails at any time, the other generator will be transferred from load 2 and take over load 1. When the prime source is restored, both loads are retransferred to the normal source and the generators shut down.

SYSTEMS AND HARDWARE

IEEE
Std 446-1980

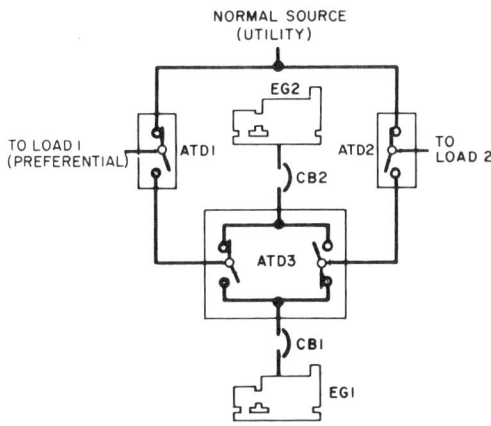

**Fig 9
Three-Source Priority Load Selection System**

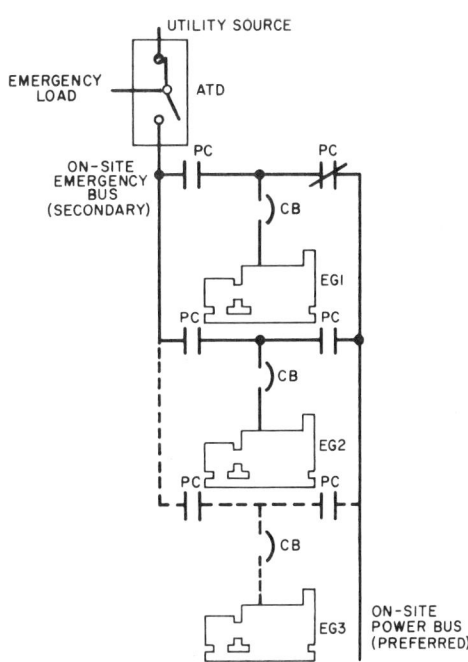

**Fig 10
Combination On-Site Power and Emergency Transfer System**

The system shown in Fig 10 provides switching and control of utility and on-site power. Two on-site buses are provided: (1) an on-site power bus (preferred) supplies continuous power for computer or other essential loads and (2) an emergency bus (secondary) supplies on-site generator power to emergency loads through an automatic transfer device if the utility service fails.

In normal operation, one of the generators is selected to supply continuous power to the preferred bus (here EG1). Simplified semiautomatic synchronizing and paralleling controls permit any of the idle generators to be started and paralleled with the running generator to alternate generators without load interruption. Anticipatory failure circuits permit load transfer to a new generator without load interruption. However, if the generator enters a critical failure mode, transfer to a new generator is made automatically with load interruption.

Many loads, such as lighting, fire alarms, heating, and air-conditioning, are fed by the utility service through the transfer device. If the utility fails, idle generators are automatically started and assume these loads through the automatic transfer device.

Fig 11 is similar to Fig 8 where idle standby generator sets can perform a secondary function by helping to supply power for peak loads. Depending on the load requirements, this system starts one or both generators to feed the peak load by switching ATD2 while the utility service continues to supply the emergency and prime loads. The second generator is paralleled automatically with the first.

If the utility service fails, the emergency and prime loads are automatically transferred to the emergency generators. Depending on generator capacity, the peak-

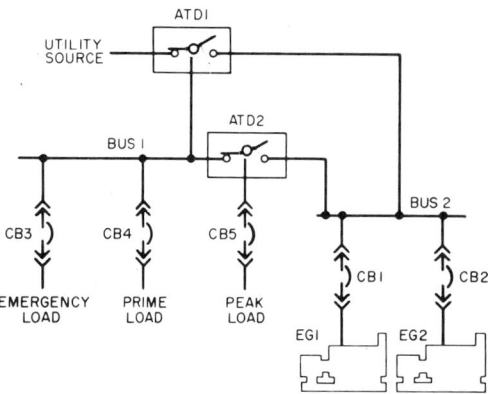

**Fig 11
Dual-Engine-Generator Standby System**

ing load may be left on, or CB5 may be dropped for the emergency.

4.2.9 *Special Considerations.* Unusual conditions of altitude, ambient temperature, or ventilation may require either a larger generator to hold down winding temperatures or special insulation to withstand higher temperatures. Generators operating in the tropics are apt to encounter excessive moisture, high temperature, fungus, vermin, etc, and may require special tropical insulation and space heaters to keep the windings dry and the insulation from deteriorating.

4.2.10 *Engine Generator Set Rating.* For some buildings the maximum continuous generator load will be the total load when all equipment in the building is operating. For others it may be more practical and economical to set up an emergency circuit or circuits so that only certain essential lights and equipment, and perhaps just one elevator, can be operated when the load is on the standby generator.

4.2.11 *Motor-Starting Considerations.* Knowing the maximum momentary voltage dip that is acceptable in the circuit, it is possible to select the size of the engine-driven generator that will be able to start given size motors without exceeding the voltage dip. If it is possible that two motors may start together, the sum of their horsepower ratings must be used as a basis for estimating motor starting requirements or controls provided for separate starting.

Engines driving generators must be sized to handle the continuous kilowatt load to be supplied to the load plus motor starting requirements and the generator losses at the number of kilowatts selected.

In sizing the engine generators for motor starting, the "locked-rotor" or inrush kilovolt-ampere rating of the motors must be used. Manufacturers' data can usually be obtained, giving the maximum rating in kilovolt-amperes of the engine generator, as well as its continuous rating. The maximum rating would be the maximum number of short-duration kilovolt-amperes available for motor starting duty without exceeding a specified voltage dip. Motor starting load has a very low power factor which must be considered in calculating the voltage dip. One other important consideration is the effect of the generator voltage dip on the motor starting torque. The starting torque is proportional to the kilovolt-ampere input to the motor, but since a voltage dip to 70 percent of rated voltage results in a reduction of power to approximately 50 percent of the number of kilovolt-amperes into the stalled rotor, and thus a 50 percent reduction of motor starting torque, problems could result in starting motors under load, unless this factor is taken into consideration.

Generators are usually sized for the maximum continuous kilovolt-ampere demand. Should there be unusually high inertia loads to start without benefit of reduced voltage starting, or if voltage and

SYSTEMS AND HARDWARE

frequency regulation other than specified cannot be tolerated during the startup period, a larger generator may be required.

4.2.12 *Load Transient Considerations.* A voltage regulator with sufficient response is required to minimize voltage sags or surges after load transients. The engine generator set must be of sufficient size and design capability to minimize the effect of load transients. Many industrial applications can tolerate large voltage sags (usually down to 80 percent, but as low as 65 percent in special cases) as long as they are not so low as to cause motor contractors to drop out or automatic brakes to set. Solid-state controls and computers will be effected.

4.2.13 *Manual Systems.* Manually controlled standby service is the simplest and lowest cost arrangement and may be satisfactory where an attendant is on duty at all times and where automatic starting and transfer of the load is not a critical requirement.

4.2.14 *Automatic Systems.* In order for engine-driven generators to provide automatic emergency power, the system must also include automatic engine starting controls, automatic battery charger, and an automatic transfer device. In most applications, the utility source is the normal source and the engine generator set provides emergency power when utility power fails. The utility power supply is monitored and engine starting is automatically initiated once there is a failure or severe voltage or frequency reduction in the normal supply. It automatically transfers the load as soon as the standby generator stabilizes at rated voltage and speed. Upon restoration of normal supply, the transfer device automatically retransfers the load and initiates engine shutdown.

4.2.15 *Automatic Transfer Devices.* Transfer equipment for use with engine generator sets is similar to that used with multiple-utility systems, except for the addition of auxiliary contacts that close when the normal source fails. These auxiliary contacts initiate the starting and stopping of the engine-driven generator. The automatic transfer device may also include accessories for automatic exercising of the engine generator set and a 5 min unloaded running time before shutdown. For additional information on automatic transfer devices refer to 4.3.

4.2.16 *Engine Generator Set Reliability.* To keep the engine in good condition, whenever it starts it should run for a sufficiently long time so that all parts reach their normal operating temperatures. In case emergency power is called for only briefly, it is desirable that the engine continue to run for about 15 min after normal power has been restored. A programmed control should be added that starts the engine once a week and operates it for a set period of time, preferably under load.

Reliability and satisfaction of the standby power supply partially depends on the engine generator installations. Some key points to consider are the following:

(1) The water supply for the engine may be disrupted by the power interruption. An isolated water supply or a closed system should be specified.

(2) Antifreeze protection may be required. A room heater, electric immersion heater, or steam jacket improves starting reliability and solves the antifreeze problem at the same time. A loss-of-heat alarm may be required.

(3) A silencer on the intake and exhaust may be required by law and will deter complaints of noise.

(4) Fuel supply systems must meet local laws, regulations, and insurance requirements. The capacity stored depends on available guaranteed quick delivery

replacement, including Sundays and holidays and under all weather conditions.

(5) Fuel supplies stored underground are desirable. Above ground, antifreeze protection may be required.

(6) Gasoline and diesel fuels deteriorate if they stand unused for a period of several months. Normal testing and running on line may be used to keep fuel fresh. Inhibitors may be added to the fuel. Fuel used for several purposes will tend to be fresh.

(7) Engine vibration transferred to the building should be dampened by rubber pads or springs, and flexible couplings should be used to fuel lines.

(8) Starting aids are of three types, of which (c) is frequently preferred:

    (a) Air heated before it reaches the cylinders

    (b) Volatile starting fluid injected into the engine air

    (c) Engine block maintained warm

4.2.17 *Air Supply and Exhaust.* Exhaust piping inside the building should be covered with gas-tight insulation to protect personnel and to reduce room temperature. The exhaust piping must be of sufficient diameter to avoid exhaust back pressure. Consideration should be given to dissipating the exhaust away from air intakes and minimizing air pollution.

Some means of providing free flow of fresh air into the generator room is necessary to keep the atmosphere comfortable for personnel and to make clean, cool air available to the engine.

4.2.18 *Noise Reduction.* Vibration must frequently be isolated from structures to reduce noise. Noise-reducing mufflers are rated according to their degree of silencing by such terms as "commercial," "moderate" or "semicritical," and "high degree" or "critical," and are usually required to meet noise standards. An intake muffler is not usually installed since an engine normally is supplied with at least one intake air cleaner that also serves as an intake silencer.

4.2.19 *Fuel Systems.* To simplify the fuel supply system, the fuel tank should be as close to the engine as possible. When gasoline or LP fuel is used, it normally cannot be stored in the same room with the engine because there is a danger of fire or fumes. However, when diesel fuel is used, it can generally be stored in the same room as the engine. The building code or fire insurance regulations should be checked to determine whether the fuel storage tank may be located beside the generator set, in an adjacent room, outside, or underground.

An engine equipped to operate on gasoline, LP gas, or diesel fuel stored in a nearby tank is a self-contained system that does not depend on outside services. It is dependable and affords independent standby protection.

4.2.20 *Governors and Regulation.* Governors are of two types, droop and isochronous. With a droop-type governor, the engine's speed is slightly higher at light loads than at heavy loads. while an isochronous governor maintains the same steady speed at any load up to full load:

$$\text{speed regulation} = \frac{\text{no load r/min} - \text{full load r/min}}{\text{full load r/min}} \times 100\%$$

A typical speed regulation for a droop-type generator is 3 percent. Thus if speed and frequency at full load are 1800 r/min and 60 Hz, at no load they will be about 1854 r/min and 61.8 Hz. A droop-type governor usually is set so that it holds the desired nominal speed at full load.

Under steady load, frequency tends to

vary slightly above and below the normal frequency setting of the governor. The extent of this variation is a measure of the stability of the governor. An isochronous governor should maintain frequency regulation within ± ¼ percent.

When load is added or removed, speed and frequency dip or rise momentarily, usually 1 to 3 s, before the governor causes the engine to settle at a steady speed at the new load.

4.2.21 *Starting Methods.* Most engine generator sets utilize a battery-powered electric motor for starting the engine. The confidence level for dependable starting is no better than the reliability of the battery and its charger. A pneumatic or hydraulic system normally is used only where starting of the electric plant is initiated manually.

4.2.22 *Lighting and Battery Charging.* Some generator room lights may be powered by the engine batteries, especially if the set is to be started manually. In addition to the battery-charging generator on the set, a separate automatic battery charger is recommended for maintaining battery charge when the generator is not running. The input of the automatic battery charger should be connected to the load side of the transfer switch.

4.2.23 *Advantages and Disadvantages of Diesel-Driven Generators.* In evaluating the merits of diesel engine versus gas turbine prime movers, the following advantages and disadvantages of each should be considered.

(1) *Fuel Supply.* Gas turbines and diesel engines can generally burn the same fuels (kerosene through #2 diesel).

(2) *Starting.* Although it may be shown that gas turbines start more reliably, both types of prime movers have highly reliable starting characteristics. However, diesel engines can start and accept full load in less than 10 s, while the gas turbines normally require 30 to 90 s depending on size.

(3) *Noise.* The gas turbine operates quieter and has less vibration than the diesel engine.

(4) *Ratings.* The gas turbine is not readily available in sizes less than 500 kW, while diesel engine units range from 15 kW on up.

(5) *Cooling.* Diesel engines in the larger sizes normally require water cooling, while gas turbines are normally air cooled.

(6) *Installation.* Gas turbines are considerably lighter and smaller in size. Turbines also require less total cooling and combustion air and produce minimal vibration. Installation costs are normally less and rooftop applications are more feasible.

(7) *Cost.* First cost for diesel engines is lower than gas turbines, but overall installed cost may become comparable due to the lower installation costs of the gas turbines.

(8) *Exercising.* The cyclic operating requirements under load are more rigid for diesel units than for gas turbines.

(9) *Maintenance.* The gas turbine is a more simple machine than a diesel engine. However, repair service for a diesel engine is more readily available than for a gas turbine.

(10) *Efficiency.* The diesel engine operates more efficiently than a gas turbine under full load. However, the reduced exercising requirement for the gas turbine normally makes the turbine the lower fuel consumer in standby applications.

(11) *Frequency Response.* The gas turbine generator is superior in full load transient frequency response.

4.2.24 *Additional Information.* Engine generator specifications are treated in

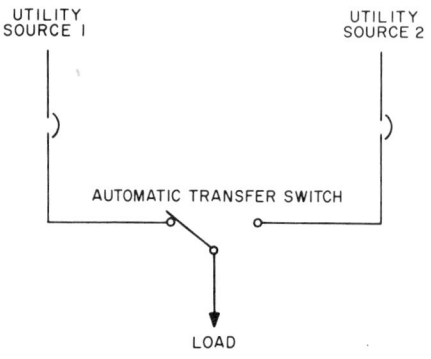

**Fig 12
Two-Utility-Source System Using One Automatic Transfer Switch**

greater detail in EGSMA GTD2-1971 and EGSMA EGS1-1970.

4.3 Multiple Utility Services

4.3.1 *Introduction.* Multiple utility services may be used as an emergency or standby source of power. Required is an additional utility service from a separate source and the required switching equipment. Fig 12 shows automatic transfer between two low-voltage utility supplies. Utility source 1 is the normal power line and utility source 2 is a separate utility supply providing emergency power. Both circuit breakers are normally closed. The load must be able to tolerate the few cycles of interruption while the automatic transfer device operates.

4.3.2 *Closed-Transition Transfer.* If the utility will permit the two sources of supply to be connected together momentarily, the transfer device should be provided with controls for both open (normal supply opened before the emergency supply is closed) and closed (emergency supply closed before the normal supply is opened) transition. With closed transition, the utility can notify the customer to transfer to the emergency source in order to take the normal supply out of service for maintenance and repair without the momentary interruption which occurs with open transition. Closed-transition requires the sources to be synchronized.

4.3.3 *Utility Services Separation.* Use of multiple utility service is economically feasible when the local utility can provide two or more service connections over separate lines and from separate supply points that are not apt to be jointly effected by system disturbances, storms, or other hazards. It has the advantage of relatively fast transfer in that there is no 5 to 15 s delay as there is when starting a standby engine generator set. A separate utility supply for an emergency should not be relied upon unless total loss of power can be tolerated on rare occasions. Otherwise, use of engine generator sets are recommended. Also, in some installations, such as hospitals, codes require on-site generators.

A no-potential alarm should be installed on the emergency supply so that the utility can be notified and emergency precautions taken if the emergency supply is lost. Additional reliability has been obtained in rare instances where the services are available from different utility companies.

4.3.4 *Simple Automatic Transfer Schemes.* Automatic switching equipment may consist of two circuit breakers interlocked as shown in Fig 13. Circuit breakers are generally used for primary switching where the voltage exceeds 600 V. They are more expensive but are safer to operate and the use of fuses for overcurrent protection is avoided. Relaying is provided to transfer the load automatically to either source if the other one fails, provided that circuit is energized. The supplying utility will normally designate

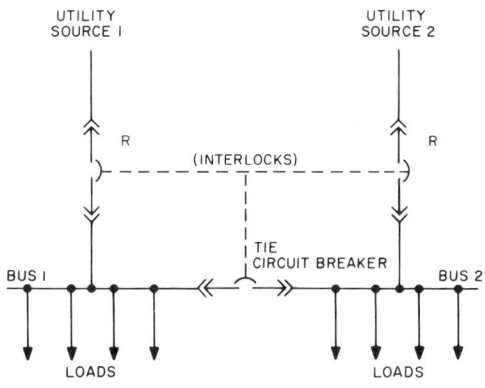

Fig 13
Two-Utility-Source System where any Two Circuit Breakers Can Be Closed

which source is for normal use and which for emergency. If either supply is not able to carry the entire load, provisions must be made to drop noncritical loads before transfer takes place. If load can be taken from both services, the two R circuit breakers are closed and the tie circuit breaker is open. The three circuit breakers are interlocked to permit any two to be closed but prevent all three from being closed. This arrangement has the advantage that the momentary transfer outage will occur only on the load supplied from the circuit which is lost. However, the supplying utility may not allow load to be taken from both sources, especially since a more expensive totalizing meter may be required. A manual override of the interlock system should be provided so that a closed transition transfer can be made if the supplying utility wants to take either line out of service for maintenance or repair and a momentary tie is permitted.

If the supplying utility will not permit power to be taken from both sources, the control system must be arranged so that the circuit breaker on the normal source is closed, the tie circuit breaker is closed, and the emergency source circuit breaker is open. If the utility will not permit dual or totalized metering, the two sources must be connected together to provide a common metering point and then connected to the distribution switchboard. In this case the tie circuit breaker can be eliminated and the two circuit breakers act as a transfer device. Under these conditions the extra cost of circuit breakers can rarely be justified.

The arrangement shown in Fig 13 only provides protection against failure of the normal utility service. Continuity of power to critical loads can also be disrupted by (1) an open circuit within the building (load side of the incoming service), (2) an overload or fault tripping out a circuit, or (3) electrical or mechanical failure of the electric power distribution system within the building. It may be desirable to locate transfer devices close to the load and have the operation of the transfer devices independent of overcurrent protection. Multiple transfer devices of lower current rating, each supplying a part of the load, may be used rather than one transfer device for the entire load.

Availability of multiple utility service systems can be improved by adding a standby engine generator set capable of supplying the more critical load. Such an arrangement, using multiple automatic transfer switches, is shown in Fig 14.

4.3.5 *Overcurrent Protection.* Caution should be exercised to assure that transfer control and operation do not in any way detract from overcurrent protection and vice versa. Furthermore, the transfer and overcurrent protective devices should be so arranged that means for disconnecting incoming service is conventional and readily accessible.

4.3.6 *Transfer Device Ratings and Ac-*

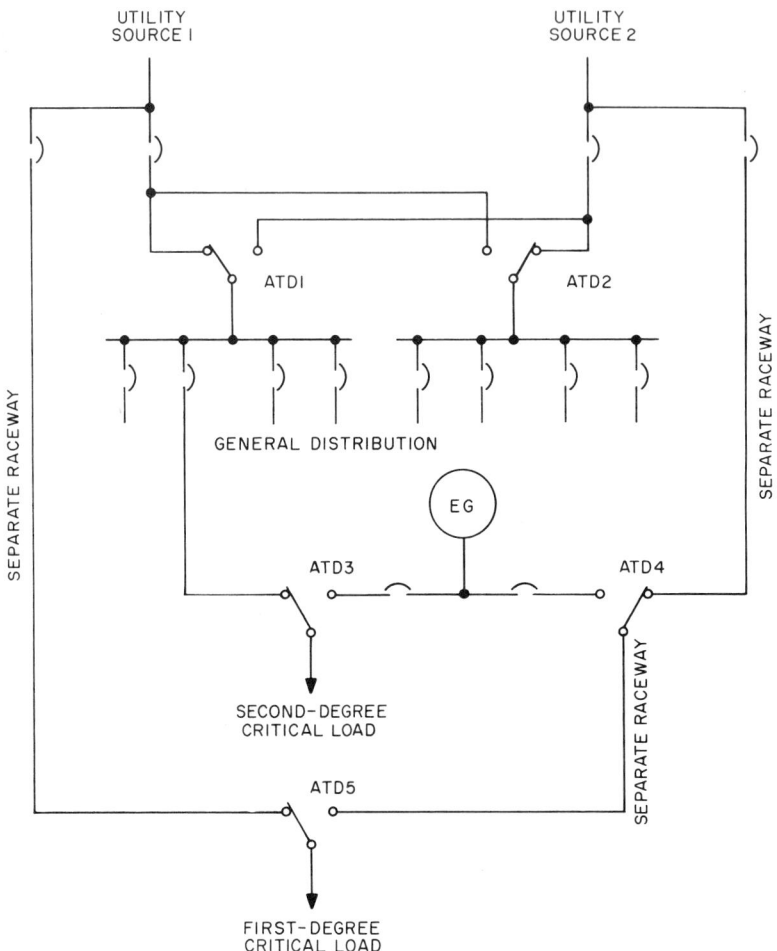

**Fig 14
Two Utility Sources Combined with an Engine Generator Set
to Provide Varying Degrees of Emergency Power**

cessories. The required characteristics of transfer devices should include their capabilities to (1) close against in-rush currents without contact welding, (2) carry full rated current continuously without overheating, (3) withstand available short-circuit currents without contact separation, and (4) properly interrupt the loads to avoid flashover between the two utility services.

In addition to considering each of the above individually, it is also necessary to consider the effect each has on the other. Particular consideration must be given to coordination between automatic transfer switches and overcurrent protection. High fault currents create electromagnetic forces within the contact structure of circuit breakers which help provide fast opening and therefore minimum clearing time. However, automatic transfer switches designed to withstand high fault current utilize these electromagnetic forces in a reverse manner to assure that the transfer switch contacts remain closed until the fault has been cleared. For these reasons transfer switching devices should be selected from those designed and approved for the purpose.

Most transfer switches are capable of carrying 100 percent rated current at an ambient temperature of 40°C. However, transfer switches incorporating integral overcurrent protective devices may be limited to a continuous load current not to exceed 80 percent of the switch rating.

Transfer switches differ from other emergency equipment in that they continuously monitor the utility source and continuously carry current to critical loads. Fault currents, repetitive switching of all types of loads, and adverse environmental conditions should not cause excessive temperature rise nor detract from reliable operation.

Most transfer switches are rated for total system transfer and thus are suitable for continuous duty control of motors, electric discharge lamps, tungsten filament lamps, and electric heating equipment, provided that the tungsten lamp load does not exceed 30 percent of the switch rating. Some manufacturers also provide transfer switches for 100 percent tungsten lamp load. There are cases where transfer switches are limited to specific loads, such as resistance loads only. For these reasons, the load classification should be determined when selecting transfer switches.

Load transfer devices are available in the following forms:

(1) Automatic transfer switches available in ratings from 30 to 4000 A to 600 V (Fig 15) [2].

(2) Automatic power circuit breakers consisting of two or more power circuit breakers which are mechanically or electrically interlocked, or both, rated 600 to 3000 A, to 15 k V.

(3) Manual transfer switches (600 V) available in current ratings from 30 to 200 A.

(4) Nonautomatic transfer switches available in ratings from 30 to 4000 A, to 600 V, manually controlled and electrically operated.

(5) Manual or electrically operated bolted pressure switches (600 V) fusible or nonfusible available from 800 to 6000 A.

Features and accessories depend upon the form of transfer device and may include the following:

(1) Undervoltage sensing monitors factory set for 95 percent pickup and 90 percent dropout (adjustable) with a time delay adjustable to 30 min after pickup will provide adequate voltage sensing for most industrial plant loads. Lower dropout or additional time delay may be necessary if significant voltage drop is produced by starting large motors.

(2) Test switch to simulate a power failure to provide a periodic test of the emergency source and the transfer operation.

(3) Controls for closed transition if the utility will permit the two sources to be momentarily tied together and the switch design permits. This will allow manual transfer from the normal to the emergency feeder and back without the momentary transfer interruption.

(4) Provisions for switching the neutral conductor and thereby minimizing ground currents and simplifying ground-fault sensing. For further discussion see Section 7.

(5) Controls for motor load transfer so as to avoid abnormal currents caused by the motor's residual voltage being out of phase with the voltage source to which

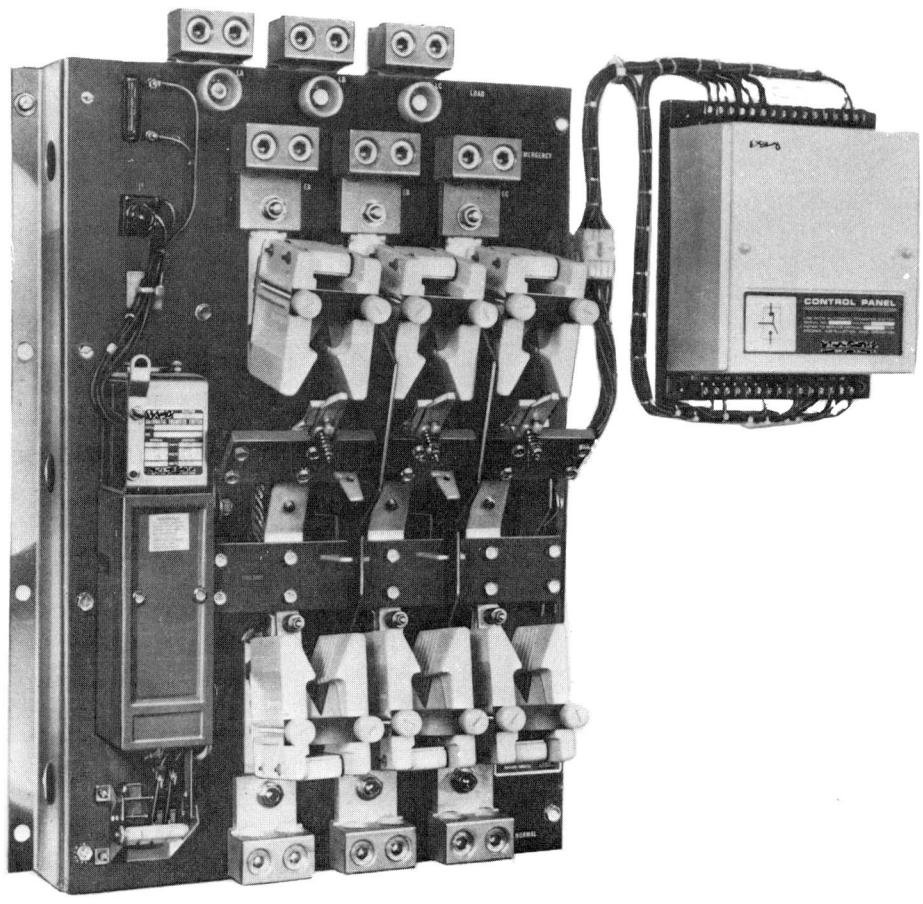

**Fig 15
Modular Type Automatic Transfer Switch Suitable for
All Classes of Load**

the motor is being transferred.

(6) Circuitry to initiate starting and stopping of the engine generator set(s) depending upon availability of the normal source of power.

Consideration should also be given to minimum voltage at which load will operate satisfactorily. The time of interruption that the load can tolerate will determine the maximum time delay permitted on the load supply devices to override momentary interruptions and, if critical, the type of transfer device.

4.3.7 *Voltage Tolerances.* The basic standard for the voltage tolerance for utilization equipment is ANSI C84.1-1977. These limits are based on the T frame motor with slightly reduced limits for equipment other than motors. In the case of special equipment, the manufacturer's tolerance limits should be obtained. Care should be taken in determin-

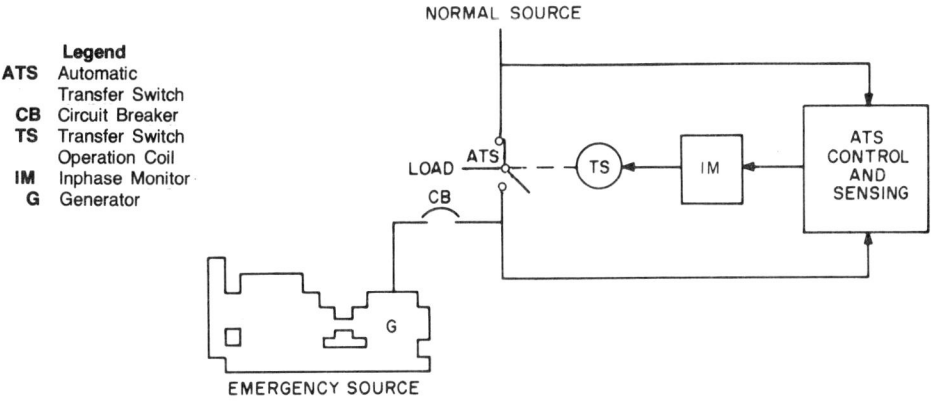

**Fig 16
Inphase Motor Load Transfer**

ing the minimum voltage for transfer switch operation to distinguish between equipment which is not damaged by low voltage, even though operation is unsatisfactory, such as incandescent lamps and resistance heaters, and equipment which will be damaged or cause damage or unsafe conditions by low voltage.

4.3.8 *Transferring Motor Loads.* Transferring motor loads between two sources requires special consideration. Although the two sources may be synchronized, the motor will tend to slow down upon loss of power and during transfer, thus causing the residual voltage of the motor to be out of phase with the oncoming source. The speed of transfer, total inertia, and motor and system characteristics are involved. On transfer, the vector difference and resulting high abnormal inrush current could cause serious damage to the motor and the excessive current drawn by the motor may trip the overcurrent protective device. Both motor loads with relatively low load inertia in relation to torque requirements, such as pumps and compressors, and large inertia loads, such as induced draft fans, etc, that keep turning near synchronous speed for a longer time after loss of power, are subject to the hazard of out-of-phase switching. Automatic transfer switches can be provided with various accessory controls to overcome this problem, including the following:

(1) Inphase transfer

(2) Motor load disconnect control circuit

(3) Transfer switch with a timed center off position

(4) Overlap transfer to momentarily parallel the power sources

Inphase transfer as shown in Fig 16 is commonly used for transferring low-slip motors driving high-inertia loads, provided that the transfer switch has a fast operating time. A primary advantage of inphase transfer is that it can permit the motor to continue to run with little disturbance to the electrical system and the process which is being controlled by the motor. Another advantage is that a standard solenoid-operated double-throw transfer switch can be used with the simple addition of an inphase monitor. The monitor samples the relative phase angle

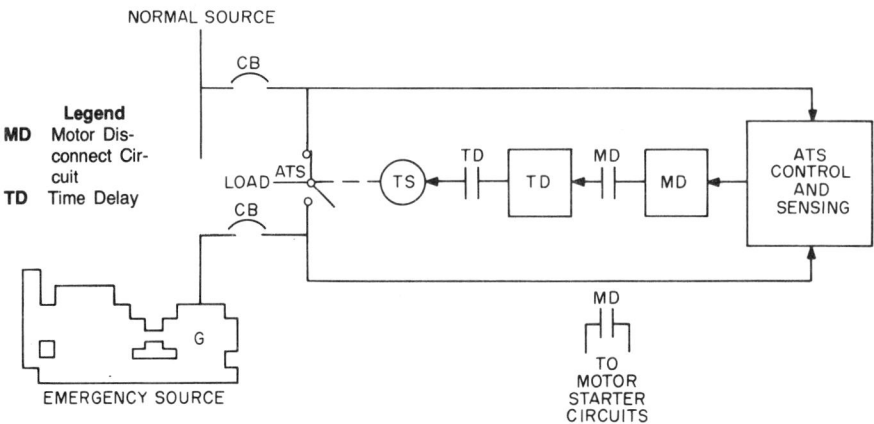

Fig 17
Motor Load Disconnect Circuit

that exists between the two sources between which the motor is transferred. When the two voltages are within the desired phase angle and approaching zero phase angle, the inphase monitor signals the transfer switch to operate and reconnection takes place within acceptable limits. Transfer switches which are equipped with inphase monitors, and which operate within 10 c (166 ms), can safely transfer motor loads without exceeding normal starting currents.

Motor load disconnect control circuits, such as shown in Fig 17 and similar relay schemes, are also a common means of transferring motor loads. However, this arrangement should not be used if the motors cannot be deenergized momentarily during transfer operations with resultant disturbances to the electrical system and the process being controlled by the motor. As Fig 17 indicates, the motor load disconnect control circuit is a pilot contact on the transfer switch which opens to deenergize the contactor coil circuit of the motor controller. After transfer, the transfer switch pilot contact closes to permit the motor controller to reclose. For these applications, the controller must reset automatically. The disconnect circuit should be arranged to open the pilot contact for approximately 3 s before transfer to the alternate power source is initiated.

Transfer switches with timed center off (neutral position) are also used for switching motor loads; Fig 18 shows a typical arrangement. One advantage, provided that timed sequence reclosing is not required, is that interconnections between the transfer switch and motor controller are not required. However, a third position (neutral) creates the possibility of the transfer switch to remain indefinitely in a neutral position owing to a control circuit or contactor malfunction. Such operation may violate NEC and UL requirements. Because there is no direct control of the motor controller, the motor controller may not drop out if it sees the residual voltage from the spinning motor. On-site tests may have to be run to establish a safe off-time under all conditions. The off-time must be long enough to permit

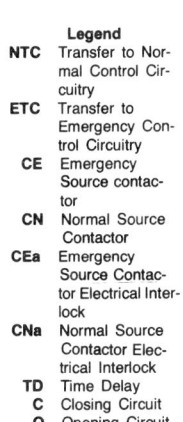
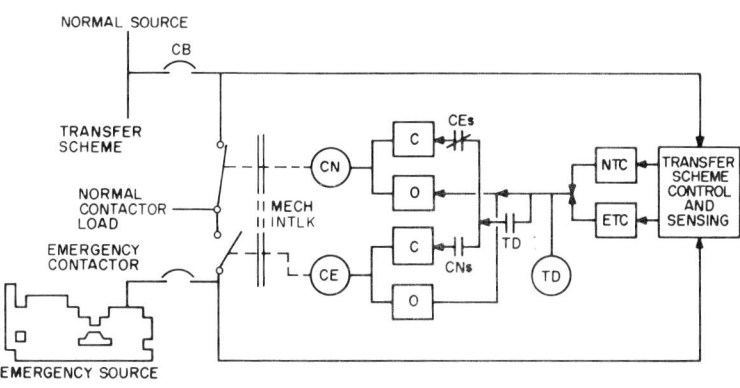

Legend
- NTC — Transfer to Normal Control Circuitry
- ETC — Transfer to Emergency Control Circuitry
- CE — Emergency Source contactor
- CN — Normal Source Contactor
- CEa — Emergency Source Contactor Electrical Interlock
- CNa — Normal Source Contactor Electrical Interlock
- TD — Time Delay
- C — Closing Circuit
- O — Opening Circuit

Fig 18
Neutral Off Position

the residual voltage to reduce to a value at which reconnection will not harm the motor, the driven load, or trip the breaker.

Overlap (closed transition) transfer with momentary paralleling of two power sources is shown in Fig 19. An uninterrupted load transfer provides the least amount of system and process disturbance. However, overlap can only be achieved when both power sources are present and properly synchronized by voltage, frequency, and phase angle. While the overlap arrangement is technically feasible, it is not always practical because of the reluctance of the utility companies to permit paralleling of extraneous power sources to their lines. However, there are indications that this reluctance is being tempered because of the energy situation.

4.3.9 *Operation of a Typical System.* Fig 20 illustrates a typical system supplying electric power to a manufacturing plant. The system is designed for initial operation with utility line 1 utilized as the normal source and utility line 2 utilized as a normally open auxiliary source. The two utility lines are synchronized with each other so that they can be paralleled, but are not normally operated in this manner. However, unless the relaying is designed for it, the operation of the two incoming lines in parallel must be kept to a minimum, that is, the switching time.

Utility lines 1 and 2 enter the plant from a substation some distance away through underground conduits separated by about 1 ft and encased in the center of a 3 by 3 ft concrete enclosure 7 ft below the surface for protection.

Operation is as follows.

(1) If voltage on the normal source (line 1) drops to 65 percent for several cycles, the undervoltage relay will deactivate, trip the circuit breaker in line 1 and close the circuit breaker in line 2 (if acceptable voltage is present on line 2).

(2) When voltage is restored to line 1, the undervoltage relay is activated and initiates a timer. After the voltage has been present on line 1 for a predetermined time (usually 1 to 10 min), the circuit

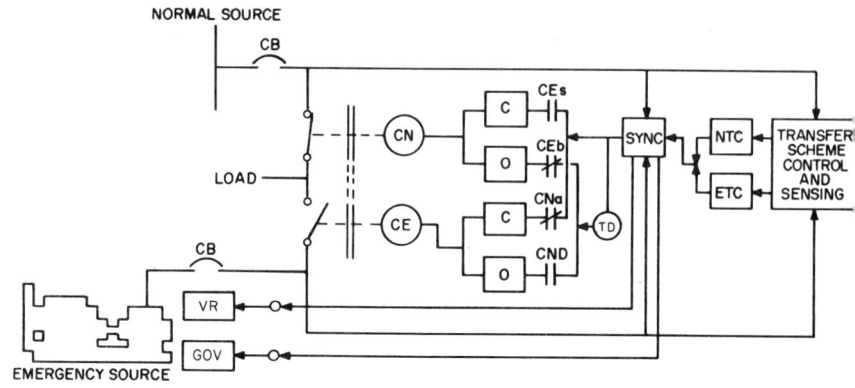

Legend
- **CEb** Emergency Source Contactor Electrical Interlock
- **CNb** Normal Source Contactor Electrical Interlock
- **TD** Time Delay to open alternate contactor if proper contactor does not operate to terminate parallel operation
- **SYNC** Automatic Synchronizer to synchronize emergency source in voltage frequency and phase angle with normal source

**Fig 19
Closed Transition Transfer**

breaker in line 1 closes, after which the circuit breaker in line 2 opens.

(3) If there is no voltage present on line 2 when line 1 loses voltage, the circuit breaker in line 2 will not close. When voltage is restored to line 1 the circuit breaker in line 1 will immediately close.

(4) If a fault or overload occurs on the load side of either incoming circuit breaker, a lockout relay keeps both circuit breakers open, disabling the automatic transfer system until manually reset.

As power demands increase, this system can be expanded by inserting a tie circuit breaker in the 13.8 kV bus and additional relaying. Part of the load would then be supplied by each utility line with transfer of the entire load to the line feeder should power be lost to one line. Load shedding of noncritical loads could be incorporated if necessary.

Operation would then be as follows.

(1) A loss of voltage or a decrease of voltage below 65 percent on either utility line will cause that normally closed circuit breaker to open and the normally open tie circuit breaker to close. When normal voltage returns, the open utility line circuit breaker will close in a preset time (1 to 10 min), after which the tie circuit breaker will open.

(2) A simultaneous loss of voltage on both utility lines will cause both normally closed utility line circuit breakers to open and the normally open tie circuit breaker to close. A return to normal voltage on either utility line will cause that utility line circuit breaker to close in its preset time and the tie circuit breaker to remain closed. When normal voltage is established on the second utility line, that utility line circuit breaker will close in its preset time after which the tie circuit breaker will open.

(3) Fault current or overload current causing either utility line circuit breaker to open will also make the automatic closing feature of the tie circuit breaker inoperative until manually reset.

(4) If the utility company needs to take one line out of service, they notify the customer who then manually closes the tie circuit breaker and opens the line to be affected.

The arrangement as shown in Fig 20 only provides protection against failure of

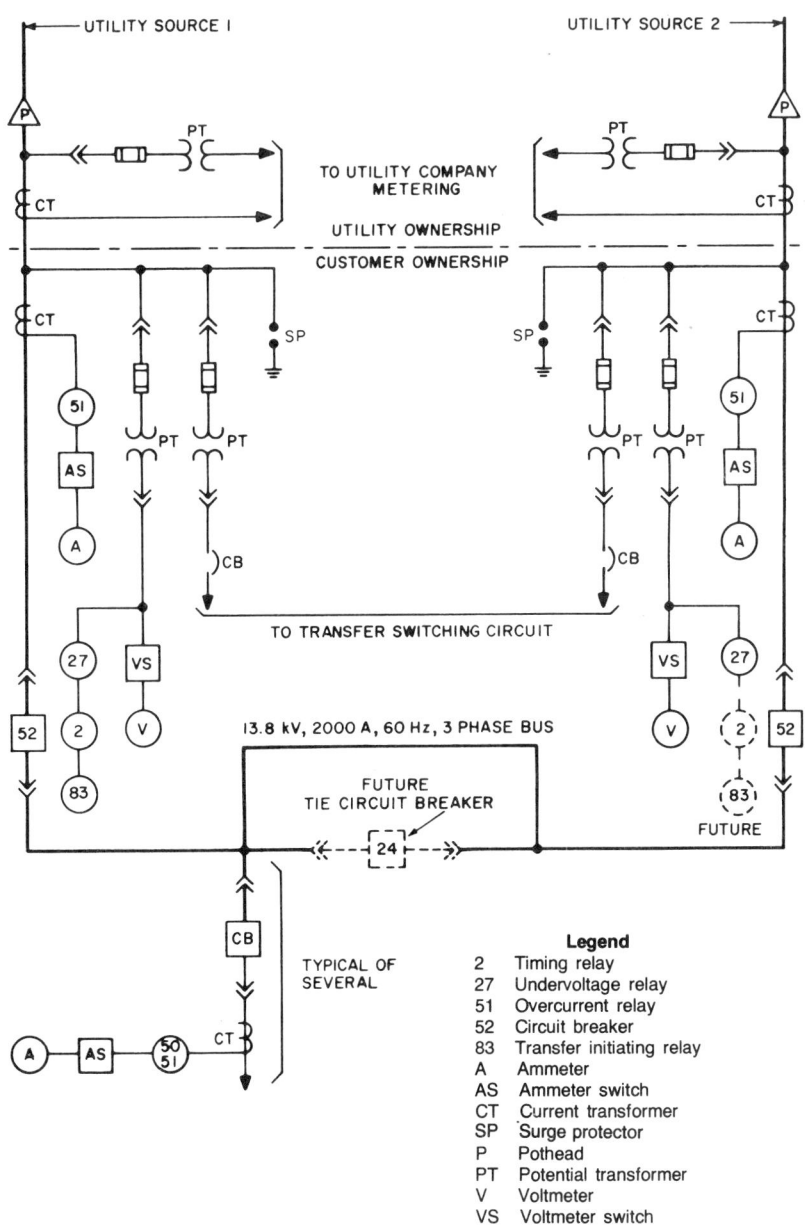

**Fig 20
Typical System Supplying Electric
Power to Manufacturing Plant**

the utility source. To provide protection against disruption of power within the building areas, it may be desirable to locate additional automatic transfer switches downstream and close to the load. A combination of circuitbreakers and downstream automatic transfer switches is shown in Fig 14.

4.3.10 *Conclusion.* Approximately 90 percent of transfer schemes designed to switch from the prime power source to the emergency source for commercial installations utilize conventional double-throw transfer switches. For maximum system reliability the transfer switches are usually located close to the load rather than at the incoming prime power source. The use of interlocked service entrance circuit breakers for transfer schemes is usually limited to medium-voltage primary switching.

Two parallel in-phase separate utility sources on line continuously with proper relaying on a normally closed tie circuit breaker provide increased reliability of electric power in some applications. This possibility should be investigated with the utility company representatives.

4.4 Turbine-Driven Generators

4.4.1 *Introduction.* Two general types of turbine prime movers for electrical generators are available, steam and gas/oil.

4.4.2 *Steam Turbine Generators.* Steam is usually not available if all electric power has been lost, although there are independent steam supply systems which themselves may have uninterruptible electrical systems. In this case steam might be considered. There are compact steam turbines which could bring power onto the line in about 5 min. This is a rather special source of supply and details are not presented.

Steam turbines are used to drive generators which are larger than those which can be driven by diesel engines. However, steam turbines are designed for continuous operation and require a boiler with fuel supply and a source of water. Thus, they are expensive for use as an emergency or standby power supply and may have environmental problems involving fuel supply, noise, combustion product output, and heating of the condensing water.

Fig 21 shows an on line steam turbine generator supplying a critical bus in parallel with one source of utility power. An alternate utility source can be manually switched on in a minute or so should there be a failure of the on line utility source. The normal utility supply system should be large enough to supply the entire critical bus if the turbine is off.

Reverse current relays should be used to immediately separate the utility supply and the steam-driven generator so that emergency power is available on the critical bus. Any protective relay scheme should be coordinated to prevent backfeed to the utility service. At the same time, two noncritical loads are shed so as not to overload the steam turbine. If the utility supply has failed, a manual transfer is made to the alternate source and the shed loads are reenergized. If the turbine has been disconnected from the bus, a decision must be made to accept a new demand peak from the utility company or to wait for the return of energy from the turbine generator before energizing the two noncritical loads. Necessary protective relaying has not been shown in the figure (see ANSI/IEEE C37.95-1973).

4.4.3 *Gas and Oil Turbine Generators.* The most common turbine-driven electric generator units employed today for emergency or standby power use gas or oil for fuel. Various grades of oil and both

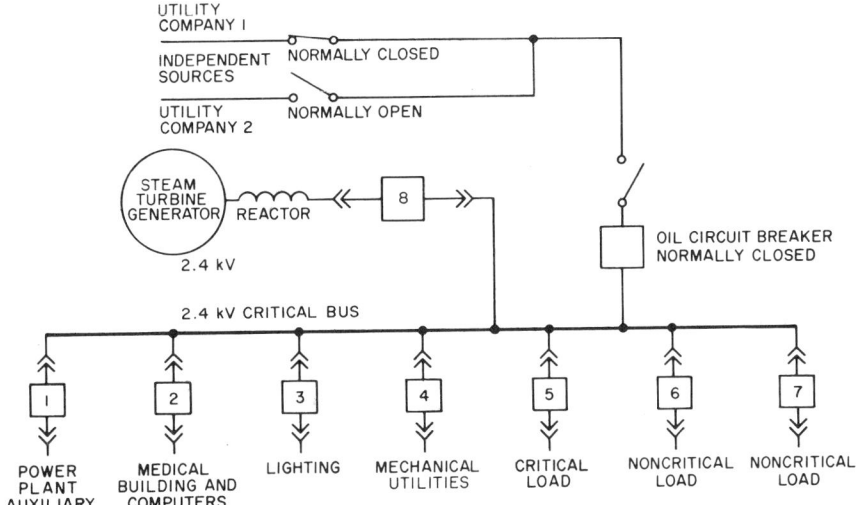

**Fig 21
Emergency and Standby Power System Using Steam-Turbine and
Dual Utility Supply**

natural and propane gas may be used. Other less common sources of fuel are kerosene or gasoline. Service can be restored from about a 40 s minimum to several minutes for larger combustion turbine units.

Aircraft-type turbines driving generators have been commonly used where electric power may be needed for a few hours to days. Small industrial units have been developed. A small fuel storage facility for safety may be adequate, provided plans have been made by which additional fuel will be delivered when needed. Care should be taken to assure an adequate gas supply should this be the source of energy, since an uninterruptible supply from a public utility company may be very expensive or unavailable. An interruptible supply may not be available when needed during cold weather. Earthquakes may destroy extensive underground utility distribution systems, but local storage may be intact. Environmental considerations may require the use of low-sulfur oil which may be difficult to obtain.

Savings are possible by the installation of a turbine for emergency and standby power when used as a peak clipping unit to reduce the demand charge. This has the operating advantage of checking the unit often enough under load so that the operating personnel are familiar with the equipment and the unit is known to be ready for an emergency. See case histories in Section 9.

Turbine designs fall between two extreme categories. Aircraft-type engines embody highly sophisticated techniques for very light weight in relation to horsepower (¼ to ½ lb/hp). This shortens life. Fig 22 shows a typical generator of this type with a riser diagram of typical loads served. Fig 23 shows a modular packaged gas turbine. The opposite philosophy em-

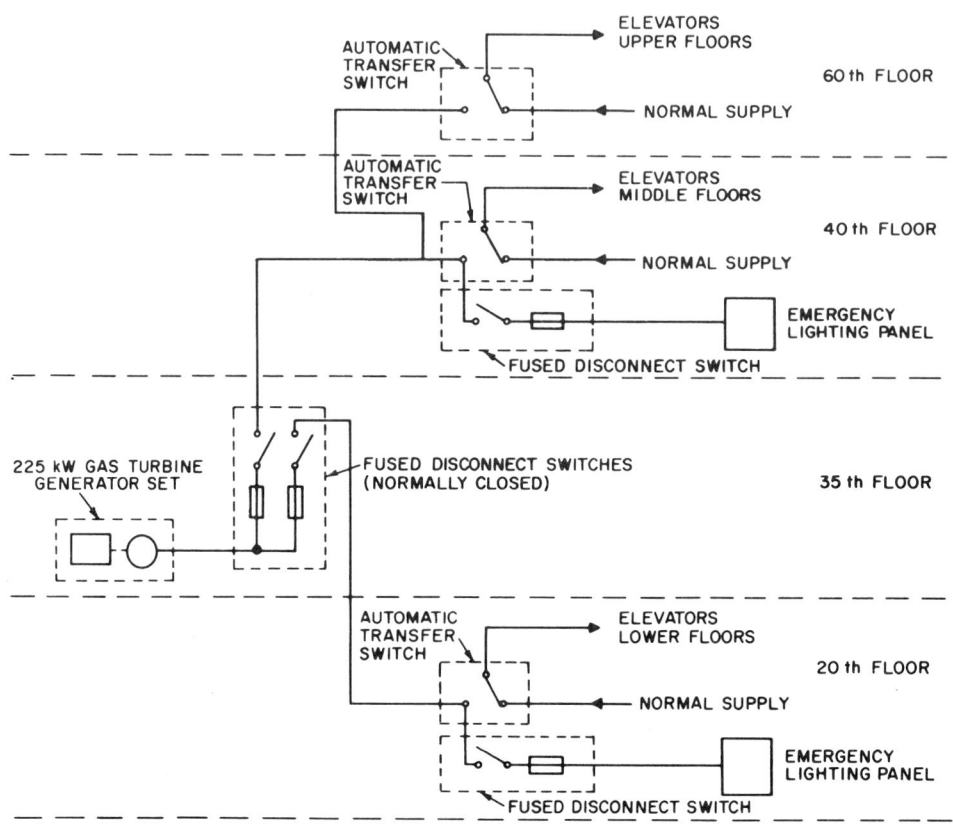

Fig 22
Typical Gas-Turbine Generator and Riser Diagram of
Auxiliary Power System

ploys the massive, bulky design techniques of steam turbines in an effort to assure long life, but with 10 lb/hp. Both have their place depending on the cost justification and hours of usage per year. Reliable industrial units are available within these ranges.

Accessory equipment for a gas turbine, such as filtering, silencing apparatus, and vibration mounts, may be required when climate conditions indicate contaminated dust-laden air or areas where noise and vibration level attenuation is required. Bands of noise from 75 to 9600 Hz are common and attenuation from 5 to 60 dB or more may be necessary in various bands.

A complete 750 kW gas turbine generator until weighs approximately 13 000 lb and occupies less than 80 sq ft of floor area. This factor of compact size and light weight can considerably reduce building costs and allow more efficient economic utilization of building space. Rooftop installations are both feasible and practicable.

The combustion turbine generator set exhibits superior performance regarding

Fig 23
Modular Packaged Gas-Turbine Generator Set Mounted on Trailer

frequency control, voltage regulation, transient response, and behavior when operated in parallel with units of the utility supply.

Combustion turbine starting and loading is accomplished either automatically or manually. Warmup is not necessary.

There are four basic starting systems available for turbines:

(1) Electric motor supplied from batteries
(2) A small steam turbine
(3) A compressed air or gas system
(4) A small diesel engine

The controls for multiple-unit installations generally involve some interconnection considerations in a master control panel or synchronizing panel. The turbines may be programmed for automatic or manual operation, start sequence, and synchronizing if desired. Special protection such as differential relays may be required.

Fig 24 shows the reduction in output capacity as gas- or oil-fired combustion turbines are installed at increasing altitudes. High air input temperatures with the associated lower air density and low barometric pressures also reduce the available output. These limitations must be taken into account or the anticipated reliability and capacity may not be obtained.

Fuel consumption at sea level will be about 14 200 to 16 250 Btu/kWh output, depending upon size and variable factors. Thermal efficiency may be raised considerably if waste heat can be used.

Installed costs will run in the vicinity of $225 per kilowatt in the medium sizes

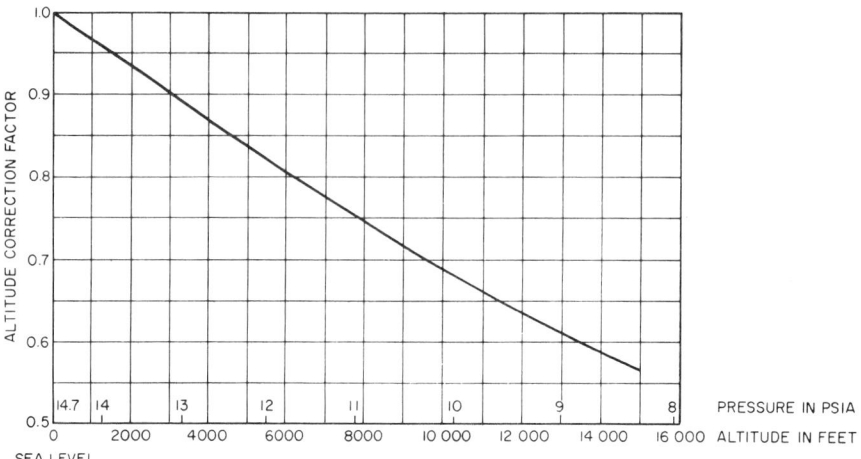

**Fig 24
Typical Performance Correction Factor for Altitude**

from 800 to 6000 kW, including fuel tank and turbine generator-package with switchgear, set in place, piped, and ducted.

4.4.4 *Advantages and Disadvantages.* See 4.2.23 for merits of diesel engines versus gas turbine prime movers.

4.5 Mobile Equipment

4.5.1 *Introduction.* An important, and yet often overlooked, source of emergency or standby power is mobile equipment. For some industrial and commercial applications, it can be the simplest, most economical, and best solution for emergency or standby needs. Mobile equipment, in the broadest sense as emergency and standby equipment, could embrace portable transformers and even substations with switching and protective devices as are available and used by most utility companies. This discussion is limited to mobile equipment as a source for electrical power. Further, in recognition that some utility company generating equipment has been located on barges and transported on waterways as needed, this discussion is further confined to land-based mobile equipment. Widest usage, by far, for most industrial and commercial applications are engine-driven and turbine-driven generating equipment. Characteristics of both of these prime movers are discussed as stationary installations in detail in 4.2 and 4.4. It is the intention of this section to pinpoint the special requirements of those prime movers when they are adapted for mobility. Some special precautions which should be followed for mobile equipment are also outlined. The decision to have mobile equipment is generally based on having multiple usages. Thus, transportability to serve several load groups, with their varying needs and electrical characteristics, will place restraints on the design/selection process of such equipment.

4.5.2 *Special Requirements.* Mobile generating units can take almost any form from only a prime mover and generator assembly to a complete self-contained plant having its own transporting power. Smaller units can be considered mobile even if they are skid mounted and must be transported by loading on and unloading from another vehicle. Obviously, there is some limit beyond which it is not practical, either because of weight or physical size to load and unload a unit for transporting, to achieve the necessary mobility. It is also obvious that a point can be reached whereby generating equipment, because of its physical size and weight, should not be considered for mobile application, even if permanently installed on a truck-bed or a trailer assembly. In terms of capacity, the size break of practicality will vary with engine or turbine drives as prime movers, equipment handling facilities, and also with how much auxiliary equipment, for support of the unit, must be transported. Based on these premises, the following is an attempt to describe the several special requirements or needs of generating equipment when it is adapted for mobile applications.

Since mobile equipment may be required to be used outdoors or, when not in use, stored outdoors, it can be desirable to house it in a weather-proof enclosure. If its use may be in a sound-sensitive area, provisions for running noise attenuation may be necessary. Whatever type of enclosure is employed, special precautions may be necessary to assure that the unit will have sufficient combustion and cooling air to operate within its rating. It may be necessary to equip the unit with louvers which are arranged to open automatically when the unit is running. For units where soundproofing is required, a well-designed forced-air ventilating system will probably be needed. For application in sound critical areas, such as residential, an appropriate high sound attenuating exhaust silencer or muffler may be warranted.

To avoid possible duplication of fuel storage facilities in all locations where the unit will find use, it may be desirable to provide fuel storage with the unit. Storage capacity will depend on each need; however, an 8 h supply as a minimum would seem to be a reasonable period within which a portable fuel transport vehicle could be dispatched to replenish the fuel supply. Fuel oil for a diesel engine is a good fuel choice because of its availability, transportability, and storability during periods of disuse and because it presents less hazards than gasoline or natural and bottled gases. Tank capacity should be based on the unit's consumption when operating at full load.

If the mobile unit is to be a towed vehicle, it will probably be required to meet all of the state's, and possibly Interstate Commerce Commission's, requirements for such things as lighting and markings. The vehicle should be designed for a realistic road speed, have stop lights, and in all probability a braking system. Thus, plug receptacle wiring connections and the hitch assembly must be compatible with similar mating equipment of the towing vehicle. Fig 25 shows a typical trailer mounted unit in the range of 15 to 45 kW.

Fig 26 shows a 2800 kW turbine-generator unit with self-contained switchgear.

A careful selection of the generator output voltage, or voltages, phases, and frequency should be made to be compatible with all envisioned load requirements. Multiwinding generators can be specified to enable field connection and selection of voltages. Most mobile rental units have multiple windings and are reconnectable for common utilization voltages.

**Fig 25
Typical Trailer Mounted Model
15 to 45 kW Capacity**

Often with smaller mobile units it is advantageous to transport the cable connection with the generating equipment. If connection time is critical in restoring electrical service, the cable connection can be hard-wired at the generator terminals, transported on an integral reel, and equipped with a plug to mate with the load-end receptacle. That receptacle could, in turn, be hard-wired into the load facility's electrical system. Many telephone building facilities and switching centers are arranged for such a plug-in connection of a mobile standby unit. The cable length should be determined for the extreme location where the equipment will be used. If a cable reel is used, the diameter should be selected after consideration of the manufacturer's recommended minimum bending radius for the cable. The cable should be of a portable variety as recognized by the National Electrical Code.

It is considered good practice to install a protective device, such as a circuit breaker, on mobile units and to connect it to the generator output terminals. Not only will the circuit breaker serve to protect the generator against overloads and short circuits at the load or in the interconnecting cable, but it will also serve as a means of disconnecting the load from the mobile unit. This is especially valuable when a cable with plug and receptacle connection is made between the unit and load facility. The circuit breaker can serve as an interlocking device to assure that it is open before the plug and receptacle connection is made. In other words, the plug and receptacle connection must be made before the generator's output breaker can be closed. This feature is considered an important safety item since an operator might be tempted to use the plug and receptacle contrary to its design by using it as a load breaking device. It is important to note that a circuit breaker selection must be based on a single output voltage of the unit. If the generator is reconnectable for several output voltages, several circuit breakers with appropriate trip ratings should be on hand and arranged for easy field installation.

Speed control through governor and fuel supply systems selection for prime movers can be a critical choice for a mobile unit if no consideration of the demands of the connected load is made. The same is true in a choice of voltage-regulating methods in combination with excitation systems for the generator. Suffice to say that a more extensive analysis, before selection, should be made for a mobile unit in comparison with a stationary installation. This is because of the several different possible demands which may be placed on a mobile unit through the more diversified demands of the loads. Improper selection of speed control or voltage control, or both, could mean either money spent wastefully or causing the unit to be unresponsive to the more critical demands of some loads.

It is generally advantageous to arrange mobile units for starting with self-contained electric storage batteries because of their transportability. A com-

SYSTEMS AND HARDWARE

IEEE
Std 446-1980

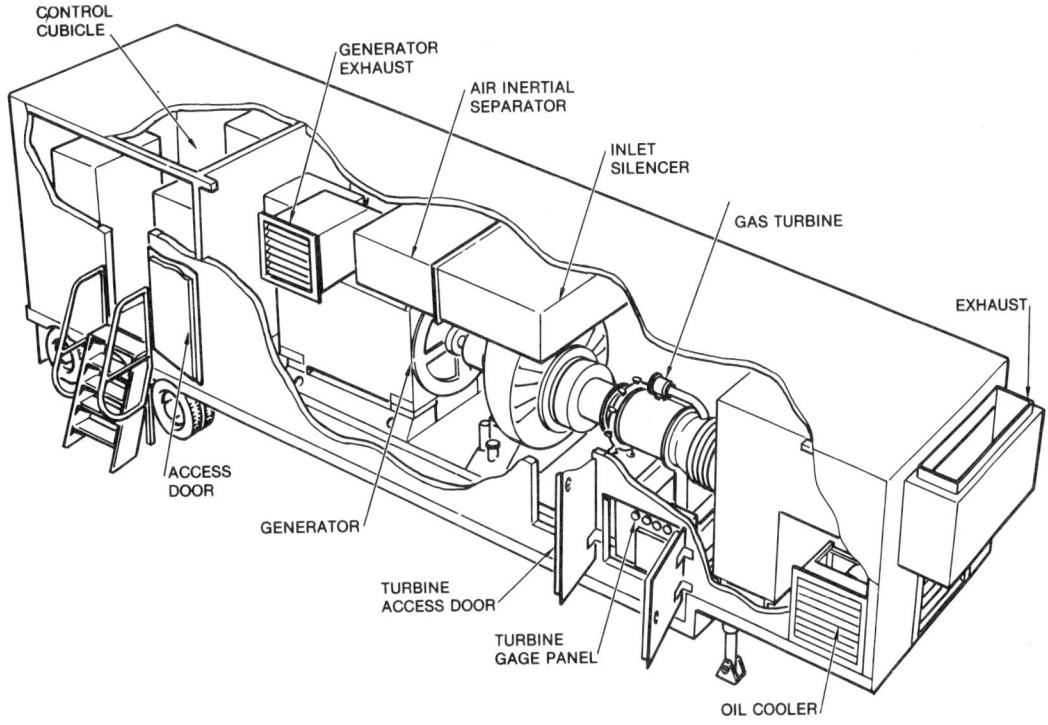

**Fig 26
Typical 2800 kW Mobile Turbine-Driven Generator Set**

pressed air start unit would require a fairly large volume air receiver which would usually reach size and weight proportions that its transportability would be difficult. If batteries are used, the permanent storage location, used when the mobile unit is not in service, should be arranged with a battery-charging unit.

If a reciprocating engine unit is to be used or stored in an outdoor location in a cold weather location, several provisions should be made, namely: use an ethylglycol solution in the radiator, where used for cooling; install an immersion heater either in the unit's crankcase or in the cooling water jacket.

4.5.3 *Special Precautions.* Mobile equipment, when compared with stationary equipment, will have some precautions which should be planned for and followed. Many of the items listed below are somewhat obvious but nevertheless are often overlooked, principally in the early stages of planning for the mobile unit's utilization and its related storage area.

If a single area is used for storing the unit when not in service, it should be equipped with a means of charging the unit's batteries through a battery charger and a flexible plug and receptacle connection. If an air start is planned, an air compressor with flexible high-pressure lines and connectors should be used. If the storage space is unheated and outdoors, in colder climates it may be advantageous to provide branch circuit wiring for connection of a unit's immersion-type heater located either in the engine's crankcase or in the unit's cooling water jacket. If there are extended periods when the unit is not used, it may be advantageous to have a dummy load bank for maintenance testing of the unit under load. Other additional features which could prove valuable for installation at the storage facility are an annunciator unit which monitors generator set key functions and an automatic exercising unit.

If a plug-in unit is to be planned, it is important that each load facility, where a mobile unit is to be used, be wired with a receptacle which mates with, and is rated for, the plug on the unit. It is often advantageous to wire the load facility with electrically or mechanically interlocked switching devices. The interlock would be arranged to prevent the possible and inadvertent paralleling of the mobile unit with the normal source of the load facility. The condition could exist when power is restored on the normal source after a programmed or unscheduled outage, during which a mobile unit was pressed into service. Because of inherent characteristics of a mobile generating unit, when compared with a facility's normal source (often a utility company), staged sequencing of returning the facility's loads to service may be in order. Rarely will automatic load sequencing be preferred over manual sequencing through a well-planned startup procedure.

Assuming that, in initial planning, it may be possible at the load facility to anticipate the operating location of mobile equipment, the fresh air intakes for an adjacent or adjunct building should be physically well-separated from the operating location for the unit.

4.5.4 *Maintenance.* Since mobile equipment's usage is often unscheduled, necessary maintenance and testing procedures can become sloppy with important items neglected. The unit should be kept ready to go at all times. If the unit is for backup of life safety equipment, it should be treated as similar, stationary, permanently installed equipment employed for the same function. As such, a regularly scheduled operating run of the unit for some representative duration under load will enable checking of such important functions as prime mover lubrication and cooling and charged batteries or starting systems, or both. This testing can often be accomplished through automatic exercising and alarm annunciation equipment installed at the storage location.

4.5.5 *Application.* Mobile equipment finds widespread application on multifaceted and physically scattered facilities such as might be operated by a municipality. Many municipals will have such diversified operations, all of which will, on occasion, require emergency and standby power, such as sewage and waste water treatment and pumping plants, garbage and refuse collection and disposal, central fire-alarm stations, centralized traffic control facilities, convention, sports, and recreational complexes, and power and steam generating plants. Other users of mobile equipment might include college and educational facilities, large governmental institutional, and industrial complexes, and military bases, all typified by scattered individual buildings, most having their own electrical service and in many cases privately owned electrical distribution facilities between buildings.

4.5.6 *Rental.* Often the expenses of purchase of a mobile unit cannot be justified when measured against how often the equipment is needed and used. There are several sources for the rental of mobile generating equipment, namely, local equipment manufacturer's suppliers and service organizations; some utility companies; and large construction con-

Table 13
Typical Rental/Purchase Costs for Gasoline-Powered Units
(8 h Cycle, 1978 Data)

Capacity (kW)	Approximate Purchase Cost ($)	Approximate Rental Cost ($)		
		Daily	Weekly	Monthly
1.5	790	12	36	110
1.75	450	13	39	120
2.0	500	14	42	130
2.5	810	15	45	130
3.0	650	17	51	150
4.0	900	21	63	190
5.0	1900	24	72	220
6.0	1900	26	78	230
7.5	2300	29	87	260
10.0	2100	35	100	320
12.5	3200	39	120	350
15	3700	48	140	430
25	5400	57	170	510
30	5400	64	190	580
35	6300	68	200	610
40	6300	75	230	680
45	6300	85	260	770

tracting firms. Table 13 shows one manufacturer's/supplier's early 1978 daily, weekly, and monthly rental rates for gasoline-powered electric plants for various sizes of units rated from 1.5 to 45 kW and based on an 8 h duty cycle. Purchase cost of each size unit is also included for comparison and are free on-board (FOB) factory for standard voltage models. In each case, listed are the costs of the basic equipments exclusive of housings, trailers, mufflers, tanks, and other similar accessory equipment. Table 14 lists similar data for diesel-powered units ranging in capacity from 3 to 50 kW operating on an 8 h duty cycle. Table 15 then lists various sizes of diesel-powered units ranging in size from 60 to 250 kW for a 12 h duty cycle. Rental equipment available on an on-call basis or from a stock basis is limited to engine-generator equipment.

Sometimes power is lost from the normal utility supply source, but a lower or higher voltage is available close by. A mobile transformer can be used for emergency power secured in time to prevent damage from freezing or in a quantity sufficient for partial production.

4.5.7 *Fuel Systems.* Growing shortages of gas and liquid fuel sources require advanced planning for a firm, readily available dependable source of supply under any emergency condition foreseen. When natural gas is planned, a firm source should be available. A back-up source of propane should be considered. Where liquid fuels are planned, readily available storage at the point of usage should be considered. Since liquid fuels deteriorate during long storage, a system of use and replacement to assure a fresh supply must be planned.

4.5.8 *Agricultural Applications.* Farm standby tractor-driven generators are available, practical, and usually reasonable in cost since the prime mover is serving a dual purpose and is usually on hand for other uses (see EGSMA TDGS1-1972 and EGSMA IMFS1-1974). See 8.3.

Table 14
Typical Rental/Purchase Costs for Diesel-Powered Units
(8 h Cycle, 1978 Data)

Capacity (kW)	Approximate Purchase Cost ($)	Approximate Rental Cost ($)		
		Daily	Weekly	Monthly
3	2200	25	75	230
5	3200	33	99	300
6	3200	39	120	350
8	3200	42	130	380
10	3200	46	140	410
12	4300	52	160	470
15	4600	57	170	510
17.5	4800	62	190	560
25	6800	65	200	590
30	6800	70	210	630
35	9100	75	230	680
40	9100	79	240	710
45	9100	86	260	780
50	9800	94	280	850

4.6 Mechanical Stored Energy Systems

4.6.1 *Introduction.* Systems under this classification deliver uninterruptible power by converting kinetic energy (KE) contained in a rotating mass to electric energy:

$$KE = \frac{(WK^2)(r/min)^2}{3.23 \times 10^6}$$

where W is the weight in pounds and K the radius of gyration in feet.

These systems provide an excellent "buffer" between the prime mover source and loads which will not tolerate transients in voltage and frequency.

By switching to standby power systems during the power support time provided by these systems, an uninterruptible power supply system may be secured for any length of time.

4.6.2 *Typical System Types.* Many practical systems are in use. Six systems will be described. Other configurations can be assembled, limited only by the

Table 15
Typical Rental/Purchase Costs for Diesel-Powered Units
(12 h Cycle, 1978 Data)

Capacity (kW)	Approximate Purchase Cost ($)	Approximate Rental Cost ($)		
		Daily	Weekly	Monthly
60	10 000	110	330	990
75	12 000	120	360	1100
100	13 000	140	430	1300
125	15 000	160	480	1400
150	18 000	190	560	1700
175	19 000	220	660	2000
200	20 000	260	780	2300
230	21 000	310	920	2700
250	23 000	350	1000	3100

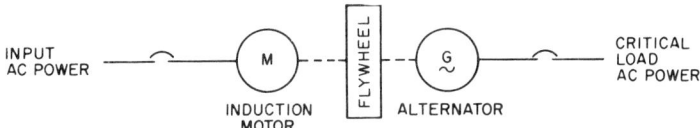

Fig 27
Simple Inertia-Driven "Ride Through" System

need, economics, and ingenuity of the designer.

(1) System 1 (Fig 27) is composed of an induction motor with low-slip characteristics which drives a flywheel and a synchronous generator. Under full rated load the output frequency is 59.8 Hz. When input power is lost, the energy stored in the flywheel drives the generator. The frequency is maintained above 59.5 Hz for time intervals of up to 0.5 s. The time interval for which frequency can be maintained is proportional to the ratio of the flywheel inertia to the load, for any given operational speed. To keep the weight of the system low, high speed is desirable. To keep noise output down and to optimize reliability, low speed is desirable. Commonly the system is operated at 1800 r/min as a good compromise speed.

Most interruptions of alternating-current input power are of short duration, lasting for intervals of from ½ to 30 cycles. Hence this system will minimize computer malfunction and loss of boiler controls from line transients at relatively low cost. However, its operational speed and hence its output frequency are directly proportional to input frequency. If a backup engine generator set is used in emergencies to power this type of critical load protection, it should have a highly accurate frequency control which will hold 60 ± 0.25 Hz for all load conditions. To do this it is necessary to provide a separate engine generator set for use only for backup to computer power in order to prevent frequency transients imposed by other loads cycling on or off, such as air-conditioning, elevators, etc. This backup generator is connected to the induction motor. The engine generator set may have to be oversized in order to start the low-slip induction motor.

Upon loss of input power, the decay of output frequency is essentially linear down to 85 percent of speed while the output voltage can be held to ±3 percent during this period.

For units up to 100 kW a minimum of 300 ms ride-through is required for a system whose minimum tolerable input frequency is 59.5 Hz. For a system which can tolerate an input frequency of 58.5 Hz, the ride-through capability would be 1.8 s.

System 1 has relatively low cost but provides minimum protection of the computer against alternating-current input power loss or transients. Medium-size units (125 kW) cost about $280 per kW.

An option available for these units senses the output frequency of the generator to initiate the shutdown sequence. This option effectively measures the energy stored in the flywheel, regardless of the sequence of voltage dips and momentary interruptions which may have reduced the speed of rotation of the set. Thus the assurance is greater that the shutdown initiating signal is given at the optimum time than is possible with input

Fig 28
Battery-Supported Inertia System

undervoltage or voltage-loss sensing and a time delay.

(2) System 2 (Figs 28 and 29) is effectively an alternating-current flywheel motor generator set with a direct-current machine and a battery bank added. In normal operation, the alternating-current motor drives the alternating current generator to supply the load. An option is available which permits the direct-current machine to act as a generator to trickle charge the batteries. Upon loss of alternating-current power, the alternating-current starter drops out and a direct-current contactor closes, applying battery voltage to the direct-current machine. The direct-current machine then operates as a motor to drive the generator. The inertia of the flywheel and the rotating machines buffer the transitions between normal and emergency operation.

When alternating-current input power is restored, the system reverts automatically to the normal mode, and the direct-current machine recharges the batteries (if that option is included). The length of interruption through which the load can

Fig 29
Diagram of Battery-Supported Inertia System

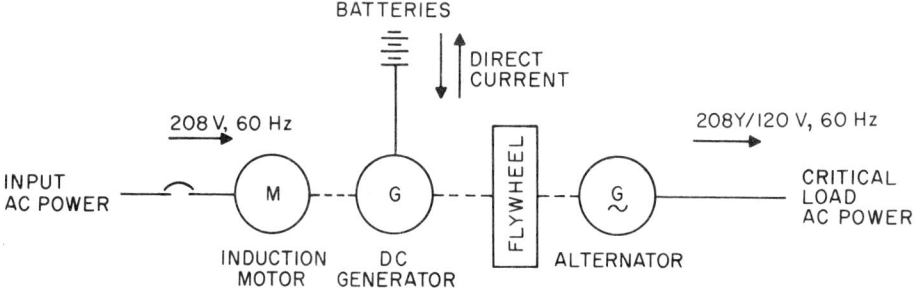

SYSTEMS AND HARDWARE

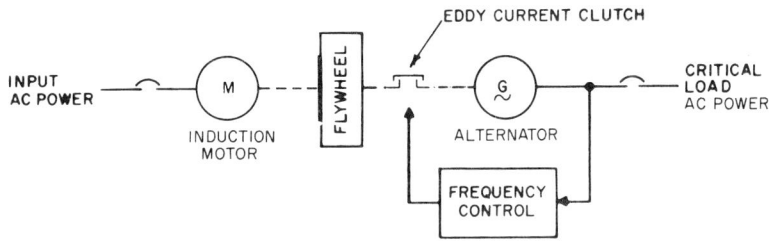

Fig 30
Constant-Frequency Inertia System

be sustained is determined by the amount of installed battery capacity.

The battery capacity may be chosen based upon the time required to start and bring on line an auxiliary generator, or an orderly shutdown may be employed.

(3) System 3 (Fig 30) consists of an induction motor which drives a flywheel and an eddy-current clutch at a fixed speed. The generator operates at a lower speed than the flywheel by controlling the slip of the eddy-current clutch, and output frequency is maintained at 60 Hz ± 0.25 Hz by this control.

On loss of alternating-current input power the generator receives energy stored in the flywheel. As the flywheel slows, the slip of the eddy-current clutch is reduced so as to maintain the 60 Hz output.

Power to the critical load is maintained for up to 15 s after loss of alternating-current input power. This provides time to start and switch in the backup power source which is normally an engine generator set with adequate power capacity to start and operate the connected loads.

Efficiencies are poor, being usually less than 55 percent at full load. Reliability is generally limited by the bearings supporting the extremely heavy and high-speed flywheel. Purchase cost is about $600 per kW for a medium-size unit (125 kW), not including installation, based on 1978 cost. For preliminary budget figures 50 percent should be added to this cost to cover the installation.

(4) System 4 is shown in Fig 31. An induction motor is driven from the utility supply and this motor is directly coupled to an alternator with its own excitation and voltage-regulating system. Coupled directly to the motor generator set is a large flywheel with one member of a magnetic clutch attached to the flywheel. The other half of the clutch is connected to a diesel engine or other prime power. During the transient period of power changeover, the kinetic energy in the flywheel is used to generate power and the voltage regulator maintains the voltage. With proper selection of components to minimize the start and runup times of the diesel engine, the frequency dip can be kept to approximately 1.5 to 2 Hz without paying a premium. Thus with a steadystate frequency of 59.5 Hz, the transient frequency would be from 57.5 to 58 Hz. The time for the diesel to start, come up to speed, and take load would normally be from 6 to 12 s.

The cost of the equipment is from $900 to $1200 per kW with 150 kW being the

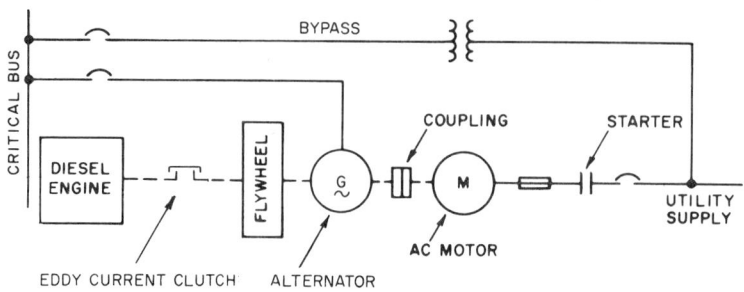

Fig 31
Rotating Flywheel No-Break System

practical limit, without sacrifice of performance or increase of cost per kW.

The advantages of system 4 are:
(a) Low first cost
(b) Moderate maintenance cost
(c) No heavy battery or battery ventilation equipment required

Its disadvantages are:
(a) Steady-state frequency to critical bus always below 60 Hz
(b) Transient frequency dip to approximately 57.5 to 58 Hz
(c) Only one chance for diesel start before frequency dips below 58 Hz
(d) Requires vibration mounts or sound insulation pads to prevent transmission of rotating equipment noise to other parts of building
(e) The diesel engine supplied with this system is limited to supplying power only for the critical bus and cannot be used to supply other standby power

(5) System 5 is shown in Fig 32. With this system, utility power is rectified to drive the direct-current motor and keep the batteries charged. The motor drives an alternator with its own excitation and voltage regulator. The alternator supplies the critical bus. Upon loss of utility power, the stored energy in the battery will keep the motor generator set running with no disturbance in frequency or voltage of the critical bus supply. The amount of stored energy will depend on battery size. A 5 min capacity is usually ample for this type of system. In order to maintain continued supply of the critical bus during prolonged interruptions, it is required to have a standby alternator set of sufficient capacity to recharge the battery and supply the full critical load. The standby alternator set must be complete with its own control and automatic transfer equipment. The size of the standby alternator will be determined by the requirements of the critical bus and battery and other essential loads.

The cost of this equipment, including a standby diesel alternator, will be approximately $1100 to $1400 per kW, with 200 kW being the practical limit.

The advantages of system 5 are:
(a) Moderate first cost
(b) Steady-state frequency is 60 Hz ± 0.25 Hz
(c) Can be used with standby diesel alternator set for other than critical bus function

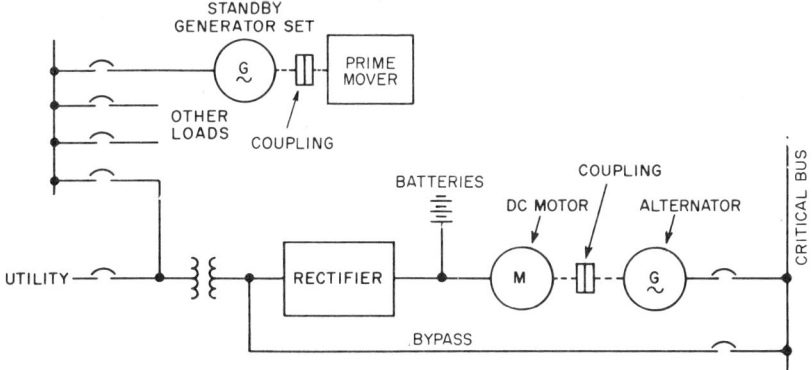

**Fig 32
Engine-Generator Supported Battery Inertia System**

Its disadvantages are:
(a) Higher operating cost
(b) Maintenance of direct-current motor commutator
(c) Maintenance and ventilation of storage batteries
(d) Heavy weight of storage batteries
(e) Heavy weight of the motor generator set

(6) System 6 (Fig 33) is a steam turbine powered no-break configuration and consists of a squirrel-cage induction motor fed from the utility supply. It is used to drive a 60 Hz alternator which is complete with its own voltage control and excitation equipment. Coupled to the motor generator set is a simple steam turbine complete with governor to actuate the throttle control valve and a diaphragm-operated main steam valve. With failure of utility power the diaphragm admits steam to the throttle valve and thus to the steam turbine which becomes the prime mover. The valve system is supplied with a bypass to allow bleeding steam through the turbine at all times to keep it hot and avoid condensation.

With this system the rotor of the turbine is rotating at all times and the output speed of the turbine shaft matches that of the generator set which is normally driving it. Since the rotating mass does not have to be accelerated when utility power fails, the frequency deviation during changeover can be quite small.

The cost of this type of set is approximately $600 per kW including all controls.

The advantages of system 6 are:
(a) Low first cost
(b) Moderate maintenance costs
(c) No heavy battery or battery ventilation equipment required

Its disadvantages are:
(a) Steady-state frequency to critical bus is always below 60 Hz
(b) Requires guaranteed supply of steam at all times to provide uninterruptible power when utility supply fails

4.6.3 *Buffer Performance.* Table 16 shows power buffer performance for three

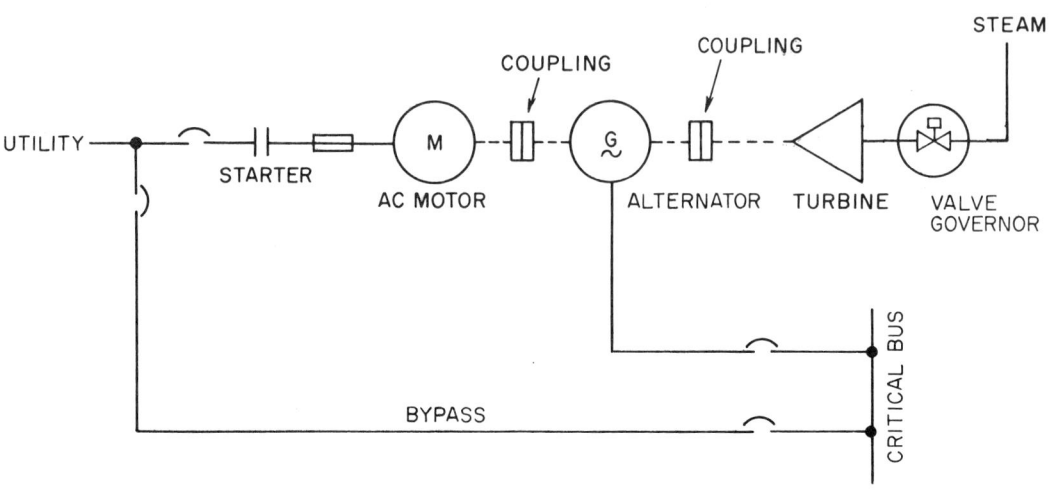

Fig 33
Steam-Turbine-Driven Emergency Power System

Table 16
Power Buffer Performance of Typical Mechanical Stored Energy Systems

	Motor Generator and Flywheel	Motor/Flywheel/Clutch/Generator	AC Motor/Flywheel Battery/DC Motor/AC Generator
Duration of emergency source	Up to 0.5 s	Up to 15 s	For length of battery supply purchased
Voltage regulation	208Y/120V ac ± 1%	208Y/120V ac ± 1%	208 Y/120 V ac ± 2%
Voltage drop or rise for 33 percent load step change from full load	± 8%	± 8%	± 10% (50% step)
Voltage transient	0.5 s	0.5 s	—
Frequency regulation	60 Hz + 0, −0.5	60 Hz ± 0.5	59.7 Hz ac drive/60 Hz ± 0.5 Hz dc drive
Frequency transient	± 0.5 Hz	± 0.5 Hz	—
Frequency transient recovery time	0.5 s	0.5 s	—
Phase angles, unbalanced loads up to 20 percent	120° ± 5°	120° + 5°	—
Harmonic voltage	5% rms maximum	5% rms maximum	3% rms maximum
Electromagnetic interference	MIL-1-16910 or better	MIL-1-16910 or better	—

SYSTEMS AND HARDWARE

typical mechanical stored energy systems.

4.7 Battery Systems

4.7.1 *Introduction.* A battery is the most dependable source available for emergency or standby power, and when applied with other devices, can also be one of the most versatile.

There are various categories of batteries, that is, automobile, aircraft, marine, motive power industrial, and stationary industrial. The stationary industrial battery is primarily used to supply backup power for communications, emergency lighting, fire alarms, switchgear, generator set cranking, etc.

When facility power is interrupted, we immediately think of people-related needs such as emergency lighting and elevator power. However, long before these would have presented a problem, many other very expensive processes may have ceased to operate. In many industrial applications a change in voltage of 15 to 20 percent or a change in frequency of only 1 percent will cause the operation to stop or perhaps even worse, if the operation continues, the product being manufactured will itself be unusable.

4.7.2 *Application Information.* Stationary batteries are configured by series connecting a group of individual cells to make up the desired voltage.

There are basically two different types of batteries used in stationary industrial applications: one is the lead-acid electrochemical couple, and the other is the nickel-cadmium electrochemical couple. There are of course variations of each type and in turn various advantages and disadvantages for the different battery types.

It can be briefly stated that the lead acid battery is less expensive to purchase than its nickel-cadmium counterpart. However, this initial capital cost may be offset in many applications because nickel-cadmium batteries generally exhibit a longer life, more rugged construction, and lower maintenance. The lower maintenance cost of nickel-cadmium batteries may be debatable due to the requirement of more batteries to obtain the necessary voltage.

The lead-acid electrochemical couple is nominally a 2 V couple. The nickel-cadmium battery is nominally a 1.2 V couple. Therefore, more nickel-cadmium cells would be utilized to configure a given battery than would be required for a lead-acid battery.

The number of cells in a battery for any specific system is a matter of adapting to suit the voltage available for charging and the voltage required at the end of the discharge period (voltage window). The most frequently used systems you will encounter and the number of cells normally applied are given in Table 17.

4.7.3 *Recharge/Equalize Charging.* In lead-acid batteries, even if the battery is not discharged, the individual cell voltages will begin to drift apart and approximately every 60 to 90 days the lower voltage cells will need to be brought back to full charge by increasing the charger voltage approximately 10 percent for 25 to 30 h. This is referred to as "equalizing" the battery. Nickel-cadmium batteries have much less self-discharge, and as a result, if the nickel-cadmium battery is not discharged with an external load, it will remain fully charged for many years at 1.4 V per cell. Therefore nickel-cadmium cells do not need to be equalized.

However, it must be understood that nickel-cadmium batteries do need the dual rate charging mode of the "float/equalize" battery charger.

Whether it is lead acid or nickel cad-

Table 17
Number of Cells for Desired Voltage

Nominal battery voltage	120	48	32	24	12
Number of lead-acid cells	60	24	16	12	6
Number of nickel-cadmium cells	92	37	24	19	10
Equalize/recharge voltage	143	58	38	30	15.5
Float voltage	129	51	34	26	13
End voltage*	105	42	27	21	10.5
Voltage window	143–105	58–42	38–27	30–21	15.5–10.5

* The end voltage is a limit imposed by the manufacturer of the electrical equipment being powered. However, as a general rule of thumb, lead-acid cells should not be discharged below 75 percent of their nominal voltage (1.5 V per cell), and the pocket plate nickel-cadmium cells should not be discharged below 50 percent of their nominal voltage (0.6 V per cell). To avoid deep discharge, most battery systems include an undervoltage relay which automatically interrupts discharge at a preset end voltage.

Note: It is not uncommon to vary the number of cells for a specific application.

mium, both batteries need the approximately 10 percent higher voltage to restore the discharged battery to a fully charged state.

A single rate float charger will adequately maintain a fully charged nickel-cadmium battery until it is discharged by an external load. However, once the battery is discharged, it will not recharge to more than about 85 percent at float voltage regardless of the current capacity of the charger. With each successive discharge, the nickel-cadmium battery in such a charging circuit may continue to lose capacity. This phenomenon has from time to time been referred to as "memory effect." It is really simply a result of inadequately recharging any battery. It is even experienced in lead-acid batteries. However, usually before the loss of capacity is noted, the lead-acid battery is destroyed by sulfation of the positive plates which is a rapid result in an undercharged lead-acid battery.

The charging rectifier, or battery charger, is a very important part of the emergency power system, and consideration should be given to redundant chargers on critical systems. A general formula for sizing the battery charger for an inverter system would be as follows:

$$\text{battery charger (amperes)} = \frac{\text{inverter output (VA)} \times 100}{\text{voltage input} \times \text{conversion efficiency}} + \frac{1.15 \times \text{battery capacity (A} \cdot \text{h)}}{\text{desired recharge time (hours)}}$$

The battery charger output must be derated for both altitude and temperature. These requirements have to be recognized since the user frequently establishes these conditions. A larger than normal rating may be required to make up the reduced capacity. A typical derating graph is shown in Fig 34.

4.7.4 Battery Sizing. To properly size any battery, the duty cycle must be defined with respect to the following:

(1) How many amperes?
(2) For how long?
(3) To what end voltage?
(4) At what temperature?

The size of the battery required depends not only on the size and duration of each load, but also on the sequence in which the loads occur.

The battery is sized to support the critical load until either (1) the critical load can be shut down in an orderly manner or (2) the utility power returns or an alter-

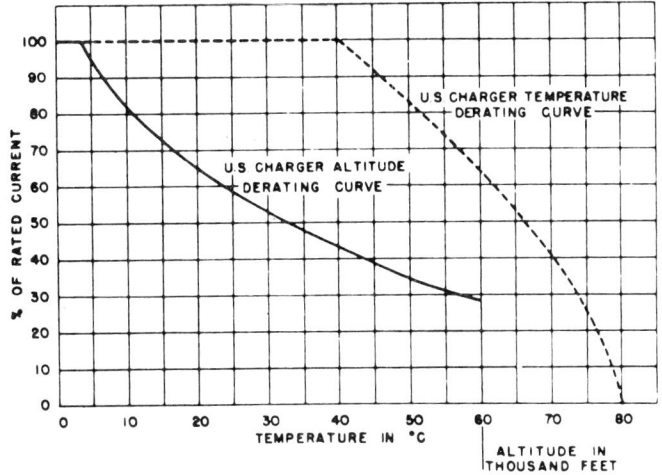

Fig 34
Derating Curves for Battery Chargers Due to Altitude and Temperature

nate standby source can be started and connected. Typical battery support times might be 5, 15, or 30 min. Rather than purchasing larger battery capacity, an engine or turbine generator standby power source should be considered.

The battery system should be sized from manufacturers' data for a particular application in a known operating temperature range. Most ampere-hour ratings are for a temperature of 77°F and a reduction of ampere-hour capacity is usually necessary for operation at lower temperatures. Some manufacturers derate their lead-acid batteries by as much as 60 percent from the 77°F ratings for operation at 0°F.

Ampere-hour (AH) capacities decrease as the rate of discharge increases. Therefore, a simple summation of the various loads (area under the current/time curve) may yield an under-sized battery.

For varying loads a summation of the various loads should be made as follows:

$$AH = A_1 T_1 + A_2 T_2 + \cdots + A_n T_n$$

where
  AH = Ampere hours
  A = Load, in amperes
  T = Time, in hours

A larger capacity battery will be required if there is a large discharge rate at or near the end of the cycle. Therefore, to verify that the battery selected is adequate, it should be checked by starting at the beginning of the discharge cycle and subtracting the energy removed AT by each load in order to determine if adequate capacity still remains for the final load interval.

Experience has shown that the lead-calcium battery requires several days to weeks to return to full equal charge on all cells following a discharge to the rated end voltage.

Other battery types may be required where frequent prolonged power outages occur, since there may not be time to fully

Table 18
General Differences for Various Battery Types

Battery Type	Physical	Typical Characteristics
Lead calcium	Pasted lead-calcium Positive plate Sulphuric acid Electrolyte	Life 12–15 years, poor in high temperatures or many or deep discharges. Lowest water loss of any lead battery. Lowest cost.
Lead antimony	Pasted lead-antimony Positive plate Sulphuric acid electrolyte	Life 10–12 years, good for cycle applications. Medium cost.
Nickel cadmium	Pocket plate construction Nickel positive plate Cadmium negative plate Potassium hydroxide Electrolyte	Life 20–23 years, good for high or low temperatures. Superior for short fast discharges. Superior for deep discharges or for many cycles. Can be rapidly recharged. Highest cost.

Note: End of battery life is defined as follows: When a rechargeable cell has been fully recharged and discharged in a load test and it fails to provide 80 percent of its original rated capacity, it has failed.

recharge between outages without raising the charging voltage beyond the rating of the connected load.

Lead planté and lead tubular batteries are also available, but their use is limited. Thus they are not included in Table 18.

ANSI/IEEE 450-1975 and IEEE Std 485-1978 provide additional guidance on age compensation and sizing, respectively.

4.7.5 *Unit Lighting Equipment.* For use in small to medium size installations, hallways, stairwells, and equipment rooms, inexpensive self-contained battery units are available. These units are available in a variety of configurations and styles, from the standard metal cabinet with attached lighting fixtures, as shown in Fig 35, to more decorative styles.

A variety of battery types are available for use in these units such as wet lead antimony, lead calcium, and pocket plate nickel cadmium. All of these require limited maintenance with the addition of water periodically. Because scheduled maintenance is difficult to ensure, the trend today is to utilize the maintenance-free type batteries, such as sealed lead calcium, pure lead, and sealed nickel cadmium.

These units are generally 6 V dc with a limited number available in 12 V dc. All of these units consist of a small automatic charger to maintain a proper charge of the battery. In the event of a power failure, the lamps are automatically switched on and supplied from the battery. When

**Fig 35
Typical Battery Unit**

normal power is restored, the lamps are turned off and the battery is automatically recharged. Performance requirements are given in NFPA No 70-1978, National Electrical Code.

Some units have provisions of powering remotely mounted lamps and exit signs, in addition to the lamps mounted on the unit itself.

Prices range from $70 to $400.

4.7.6 *Central Battery Lighting Systems.* Building wide systems may also be used to connect many lamps to a central battery charger console. The decision may be to install many 6 V units or a single 12 V, 32 V, or 115 V system with a power source (battery, charger, console) centrally located. The trend today, particularly in new construction where 10 to 100 lamps are involved, is to utilize the central system. The advantages of such a system are the following:

(1) Centralized power source, eliminating the need for single units located throughout the building, use of less space (only lamps are in the areas to be protected), and remote location of the power source. This facilitates maintenance and testing of the system.

(2) Availability of alarm and protection circuits, which increases the flexibility of the system.

(3) Advantage of distributing at lower current, higher voltage, resulting in decreased losses.

(4) Central location, which reduces maintenance costs and provides a means for regular testing of the system for reliability.

When systems larger than the 32 V or 115 V are justified, the inverter system is recommended.

4.7.7 *Factors to Consider When Selecting Emergency Lighting Systems.* Immediate questions are what area is to be protected and what light level is necessary. The layout of the building, department, and usage of rooms will answer the questions. Areas which should be lighted are places near moving machinery, passageways, exits, stairwells, etc. The light level required is dictated by several factors, including applicable codes, usage, available funds, and personal preference. The minimum light level, where required by NFPA 101-1976, is 1.0 foot-candle. The light level is determined by the number of lamps, their wattage, the lumen output of the lamps per watt, the reflector efficiency, the reflectance of the building surfaces, and the area covered.

As the number of lamps, length of wire runs, and size of lamps increase, the voltage of the system should be increased to obtain a minimum cost and a more efficient system. The 32 V system should be applied on intermediate size applications; larger applications may require the 115 V system or inverter system.

In summary, there are three steps in sizing an emergency lighting system: (1) determine number, location, and type of lamps needed; (2) determine total lamp wattage of all selected lamps; and (3) determine protection time required.

Larger systems require a distribution panel. Its purpose is to distribute the available dc power to a number of individually fused circuits with each fuse preferably monitored by a visual and audible alarm circuit. If one circuit suffers a fault, the fuse protection removes the faulted circuit from the direct current bus, allowing the battery to power the remaining circuits. The distribution panel is usually located in close proximity to the battery supply.

The normal alternating current voltage supply to the battery charger and the battery voltage should be continually monitored. An alarm should be sounded whenever supply voltage fails and if the battery voltage should become high or low. Permanent battery damage can result from continued operation with the plate surfaces exposed to the air. Therefore, an alarm system to monitor electrolyte level in the battery should be considered. The battery system alarm may be in any convenient location.

4.7.8 *Multiple Sources Used for Normal Lighting.* The systems described up to now are used whenever a single voltage source is used for normal lighting. The load contactor senses the failure of the alternating current voltage it monitors, and all emergency lamps will come on.

There are systems that will monitor each line-to-neutral voltage of a three-phase system. Thus, upon failure of any one or all of the three voltages, the load contactor will close, turning on all emergency lights.

If it is not desirable to turn on all the emergency lights, but only the room or zone where the normal lighting has failed, a multiload contactor panel may be used to monitor the alternating current voltage that supplies the lighting in the area or zone protected by that load contactor. If that voltage should fail, then that contactor only will close, and its lights will come on. If more than one ac voltage fails, then more than one string of emergency lights will come on.

When the normal lighting is supplied by mercury vapor or other high intensity discharge (HID) lamps, full light is not available during the ignition period. This period could range from 1 to 20 min, depending on temperature, line voltage, and the type of lamp. The normal characteristic of the emergency lighting system is to supply lighting only when the normal power is not available; as soon as it returns, the emergency lamps will go out. Thus there will be a period of time that no lighting will be available. To overcome this defect, a time delay on the load contactor should be considered.

The battery will try to continue to supply power as long as the normal power is not available. In actual practice, however, the battery is limited to the amount of power it can deliver. A power failure of long duration will result in an overdischarged battery. Nickel-cadmium batteries are not affected by this overdischarge, but a lead-acid battery may have permanent damage as a result of this overdischarge. In this case, lead-acid batteries should be protected by utilizing a dropout relay. This relay is sensitive to voltage and disconnects the battery once its voltage reaches a critical point, generally 87½ percent of system voltage.

4.8 Battery/Inverter Systems

4.8.1 *Introduction* Continuous and disturbance-free ac power is required for a growing number of critical load applications. The foremost example of this need is sensitive electronic equipment used for data processing, life support, communication, and control functions (see 3.10 through 3.13).

Combining the energy storage capability of batteries with solid-state inverter technology provides a reliable, high-quality ac power system. Because of the capability to operate continuously on-line, affording no break in power when the primary source fails, these systems have been designated as uninterruptible power supplies (UPS).

4.8.2 *Battery/Inverter Supply Used as Standby Source.* Fig 36 shows a relay transfer system for supplying an emergency power source to a load upon the loss of the prime source. Power to the load will be interrupted, as shown in Fig 37, from 60 to 190 ms, depending upon the type and size of the dc contactor and

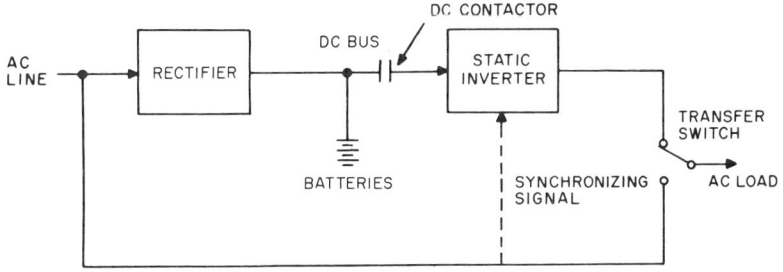

Fig 36
Short-Interruption Standby System

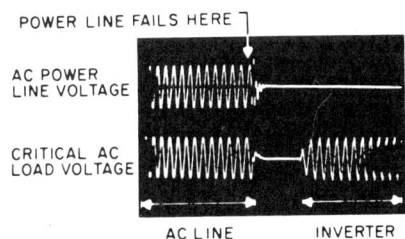

Fig 37
Oscillogram of Output Voltage of
System of Fig 36 During Transfer

transfer switch. More costly but faster transfer systems use static switches.

In the system of Fig 36 the loss of voltage on the ac line will cause the dc contactor to close and supply power to the inverter. At the same time the transfer switch operates to transfer the load to the inverter supply. This system is adequate for lighting, signal circuits, radio systems, and other loads which can tolerate the short interruption of power.

4.8.3 *Nonredundant Uninterruptible Power Supply.* The basic UPS configuration consists of a single rectifier, battery, and inverter operating continuously in the powerline. These are available in sizes ranging from 250 VA to over 500 kVA. A typical one-line diagram is shown in Fig 38.

During normal operation, the prime power and rectifier supply power to the inverter and also charge the battery which is "floated" on the direct-current bus and kept fully charged. The inverter converts battery power from direct to alternating current for use by the critical loads. The inverter alone governs the characteristics of the alternating-current

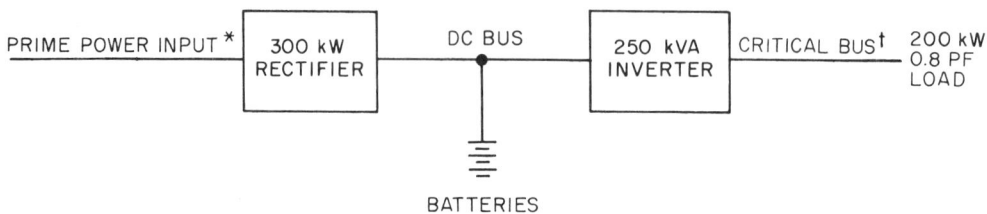

**Fig 38
Nonredundant Uninterruptible Power Supply**

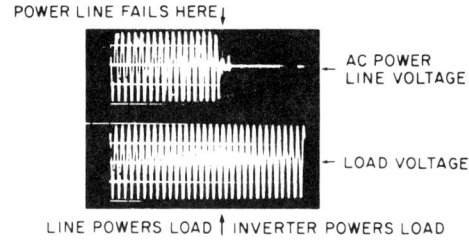

**Fig 39
Oscillogram of System of Fig 38 with
Power Line Failure**

output, and any voltage or frequency fluctuations or transients present on the utility power system are completely isolated from the critical load.

In the event of a momentary or prolonged loss of power, the battery (which is floated on the dc bus) will supply sufficient power to the inverter to maintain its output for a specified time for a few minutes to several hours until the battery has discharged to a predetermined minimum voltage.

Fig 39 shows an oscillogram of the continuous-load voltage supplied by the system of Fig 38, even when the prime source of power fails.

Upon restoration of the prime power, the rectifier section will again resume feeding power to the inverter and will simultaneously recharge the battery. The rectifier is therefore sized 1.2 to 1.5 times larger than the inverter kilowatt output rating to account for this battery charging demand and inverter losses.

Static systems, as described, provide

(1) Precise uninterruptible power
(2) Low maintenance, no moving parts
(3) High efficiency, static conversion devices

System availability should be as high as economically justifiable and may be calculated by using the following formula with all figures in the same units, usually hours:

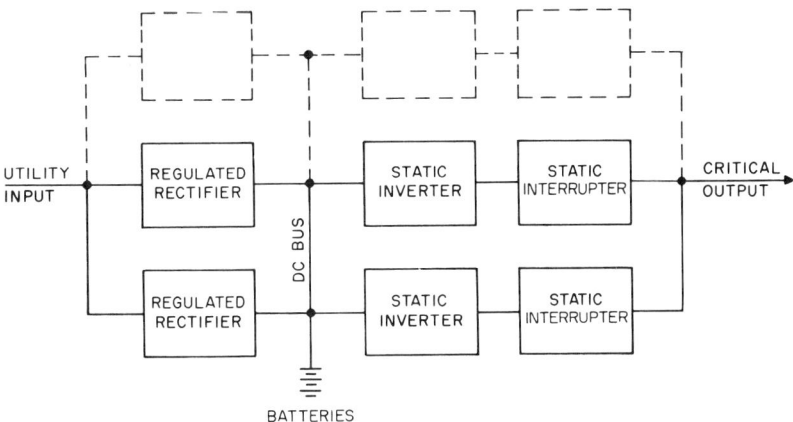

Fig 40
Redundant Uninterruptible Power Supply

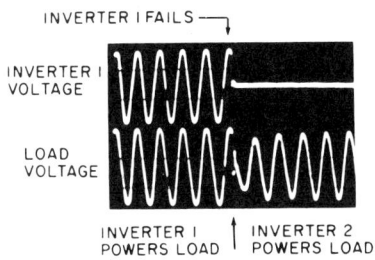

Fig 41
Oscillogram of Output Voltage of System of Fig 40 upon Inverter Failure

$$A = \frac{MTBF}{MTBF + MTTR}$$

where
  A = System availability
  MTBF = Mean time between failures
  MTTR = Mean time to repair

A typical specification for an uninterruptible power supply system is given in Table 19.

The nonredundant system shown in Fig 38 has the advantage of simplicity and low cost. Large systems can have a predicted reliability of 20 000 h of mean time between failures using handbook reliability data; but field experience indicates that reliabilities in the order of 40 000 h of mean time between failures can be achieved. The nonredundant system has the disadvantage of disturbing the critical bus in the event of an inverter failure. This disadvantage can be overcome by the use of a static bypass switch as shown in

Table 19
Typical Nonredundant 3φ UPS Performance Specifications

Input (Rectifier/Charger)
  Voltage   208 V or 480 V, ±10%, 3 phase
  Power factor   Minimum 0.8 at rated load
  Frequency   50 or 60 Hz, ±5%
  Harmonic content of current   10% (5% preferred)
  Startup current limiting   Maximum 25% of full load current (energizing rectifier transformer with inverter at no load)
  Startup "walk in"   15 to 30 s to full load
  Steady-state current limiting   Adjustable, with two standard settings:
                                             1) For utility power, 125% rated load
                                             2) For emergency power, 100% rated load plus 5 kVA

Output (Inverter)
  Voltage   208 V or 480 V, 3 phase, 3 or 4 wire
    Regulation
                                             1) ±2% for balanced load
                                             2) ±3% for 20% unbalanced load (100%, 80%, 80% or 100%, 100%, 80%)
    Line drop compensation   0 to 5%, adjustable
    Transient response
                                             1) ±5% for loss or return of ac input power
                                             2) ±8% for 50% load step
                                             3) ±10% for bypass or return from bypass
    Transient recovery   Return to steady-state conditions within 100 ms after a disturbance
    Harmonic content of voltage   4% total, 3% any single harmonic
    Phase displacement
                                             1) 120° ± 1° for balanced load
                                             2) 120° ± 3° for 20% unbalanced load
  Frequency   50 or 60 Hz
    Regulation   ± 0.1 Hz
    Line sync range   ± 0.5 to 1.0 Hz, adjustable
    Slew rate   Maximum 1 Hz/s
  Current capability
    Overload   125% for 10 m and 150% for 10 s
    Fault clearing   150% to 300% for 10 cycles, maximum limited for self-protection

DC Link (Battery)
  Battery type   Lead acid or nickel cadmium (NICAD)
  Float voltage   Lead acid 2.2–2.25 V/cell
                                             NICAD   1.4–1.42 V/cell
  Equalize voltage   Lead acid 2.35 V/cell
                                             NICAD   1.6 V/cell
  End voltage   Lead acid, minimum 1.6 V/cell
                                             NICAD   minimum 1.1 V/cell
                                                (setting also determined by inverter input voltage window)
  Recharge time   10 times discharge time
  Energy storage capacity   Sized to requirement (normally 15 min)

General Characteristics and Requirements
  3φ Output ratings   32.5 to 600 kVA at 0.8 power factor
  Efficiency   77% to 90% (improves as kVA rating increases)

Table 19 (continued)

Dimensions and weight	Depends on kVA rating
Controls	Startup, emergency shutdown, synchronous transfer to bypass and all adjustment functions required for operation and maintenance
Meters	AC volt and ammeters with phase selector switches for both input and output, DC voltmeter and charge/discharge ammeter
Alarms	Indicating 10 to 20 special conditions or malfunctions such as output over- and undervoltage, battery discharge, fan failure, auto bypass, etc
Environmental	
Ambient temperature	Within 0° to 40°C operating and −20° to 70°C nonoperating
Relative humidity	0 to 95% at any operating temperature
Reliability	MTBF 200 000 h minimum (includes available utility power via bypass)
Maintainability	MTTR 40 min maximum (when parts are on site)
Available Options	
Frequency conversion	50 to 60 Hz or 60 to 50 Hz (only for redundant type UPS without bypass)
Expandability	Can be paralleled with like UPS modules
Electromagnetic interference suppression	Suppression of radiated on all sides and conducted on input, output, and control cables
Acoustical noise suppression	Maximum 76 dB at 5 ft from surface
Extended operating temperature capability	From 40°C to 50°C
Automatic battery equalizing charge	Activated and timed after each battery discharge
Circuit breaker motor operators	For input, output, and battery circuit breakers
Mimic bus	An illuminated one-line diagram indicating operational status
Remote status monitoring and alarm panel	Monitors special conditions and malfunctions up to 500 ft away
Additional meters	Input and output wattmeters, elapsed time and frequency meters rectifier output dc ammeter
Special conditions to be identified by user	Damaging fumes Excessive moisture Excessive dust Abrasive dust Steam Oil vapor Explosive mixtures of dust or gases Salt air Abnormal vibration, shocks, or tilting Weather or dripping water Special transportation or storage conditions (user to identify method of handling equipment) Extreme or sudden changes in temperature Unusual space and weight limitations Unusual operating duty Unusually high system impedance Seismic considerations Electromagnetic fields Radioactive levels above natural background Abnormally high system voltages to ground Nonlinear load or one generating excessive harmonic or ripple current Inability for the dc source to accept a current in the reverse direction Acoustical noise limitations Type of battery or power supply provided by user

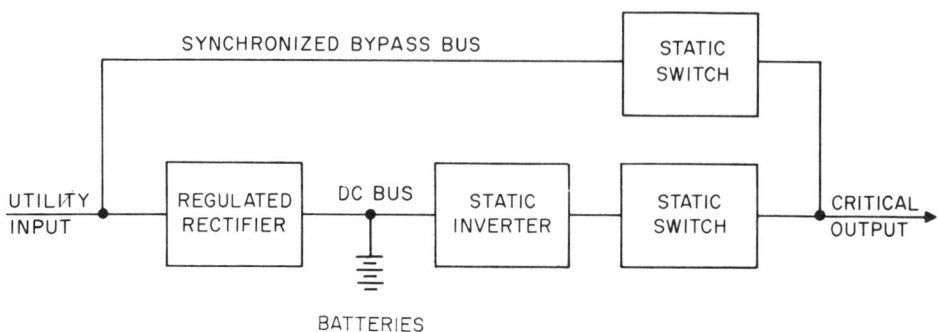

Fig 42
Uninterruptible Power Supply with Static Bypass

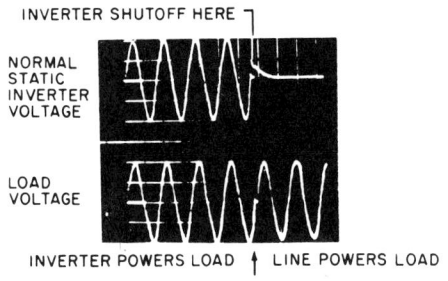

Fig 43
Oscillogram of Static Switch of
System of Fig 42 Load Voltage

Fig 42 or a redundant system as shown in Fig 40.

4.8.4 *Redundant Uninterruptible Power Supply*. Fig 41 is an oscillogram of the two-inverter transfer system voltage during malfunction of one inverter similar to the arrangement shown in Fig 40. In the redundant system, each half of the system has a rating equal to the critical load requirements. The basic power elements (rectifier, inverter, and interrupter) are duplicated, but it is usually not necessary to duplicate the battery since its inherent reliability is extremely high. Certain control elements such as the frequency oscillator may also be duplicated.

The static interrupters isolate the faulty inverter from the critical bus and prevent the initial failure from starting a "chain reaction" which might cause the remaining inverter to fail.

Large systems which require multiple rectifier/inverters to handle the load requirements would require one additional rectifier/inverter to provide redundancy. Redundant systems should consist of the fewest parallel paths required to supply the critical load requirements plus one additional path for redundancy. A larger number of smaller rated paths does not necessarily provide increased reliability since they add unnecessary additional components which are themselves subject to failure.

The cost of a redundant system is approximately $(N + 1)/N$ greater than for a nonredundant system, where N equals the least number of parallel paths required for a nonredundant system. However, such a system is two to four times more reliable than a nonredundant system.

4.8.5 *Nonredundant Uninterruptible Power Supply with Static Bypass Switch.* An alternate method of increasing the overall system reliability is a static bypass around the faulted inverter as shown in Fig 42. When an inverter fault is sensed, the critical load can be transferred to the bypass circuit in less than 5 ms. Fig 43 is an oscillogram of the load voltage as supplied during a transfer of the power source.

The static bypass adds about 10 percent to the cost of a nonredundant system, but is eight to ten times more reliable.

4.8.6 *Parallel Redundant Uninterruptible Power Supply.* Fig 44 shows a parallel supplied parallel redundant UPS system. Reliability is a paramount consideration in this 1000 kVA system. Solid-state sensors and static switches, not shown, are installed to clear a malfunctioning inverter without effect on critical computer load.

This parallel "load-sharing" configuration is usually specified when a static bypass cannot be effectively applied, such as when the utility input power quality is so poor that it cannot be relied upon for even short periods or when the utility input power is at a frequency different than that required by the load.

4.8.7 *Cold Standby Redundant Power Supply.* This configuration consists of two basic UPS systems, two static switches, and a common battery; see Fig 45.

One UPS operates on line and the other is turned off. Should the operating UPS fail, its static bypass circuit will automatically transfer the critical load to the utility source without interruption of the load. The second UPS is manually energized and placed in the bypass mode of operation. To transfer the critical load to the second UPS external make-before-break, nonautomatic circuit breakers are operated, placing the critical load on the second UPS bypass. Finally, the critical load is returned from the bypass to the second UPS inverter via the static switch. The two inverters cannot be operated in parallel and, therefore, an interlock circuit must be provided to prevent this condition.

The cold standby configuration exposes the critical load to the utility line for short periods during transfer. Nevertheless, this configuration is found to be more reliable than the parallel redundant because the inverters are operated independently and single-point failures of load-sharing control circuits are eliminated. It is also more efficient to operate since one rectifier and inverter at 100 percent load is several percent higher in efficiency than two at 50 percent load.

Cold standby redundant costs about 5 percent more than the two-module parallel redundant configuration owing to the need for two static switches. However, this higher initial investment can be justified based on lower operating costs.

The cold standby redundant is typically specified instead of a nonredundant UPS system when the installation site is isolated and logistic support is poor. Of course the steady-state utility power must be acceptable to the critical load for short periods during transfer between modules.

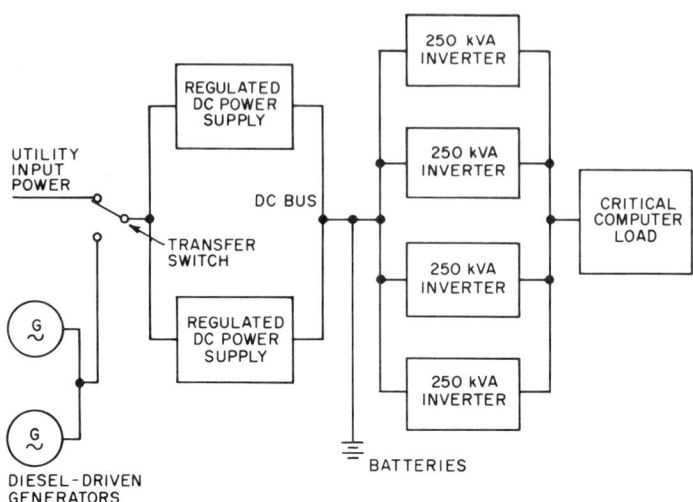

Fig 44
Parallel-Supplied, Parallel-Redundant Uninterruptible
Power Supply

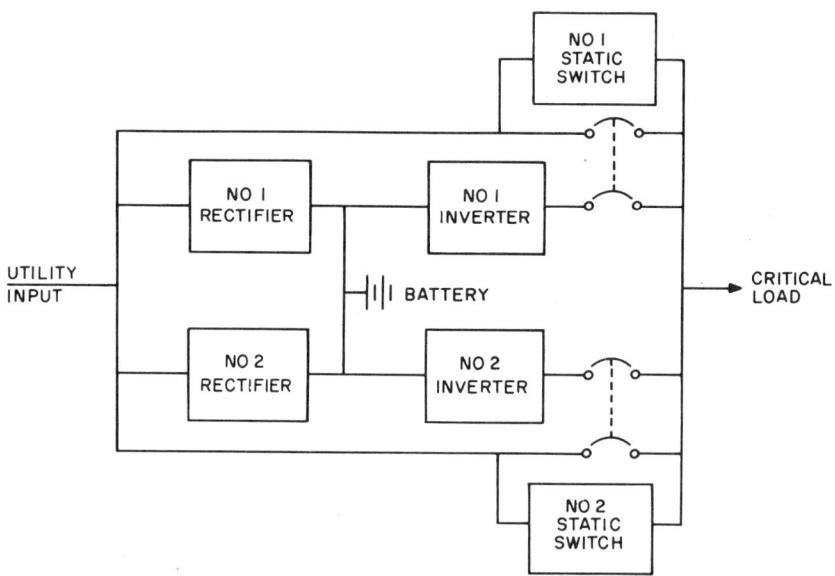

Fig 45
Cold Standby Redundant Uninterruptible Power Supply

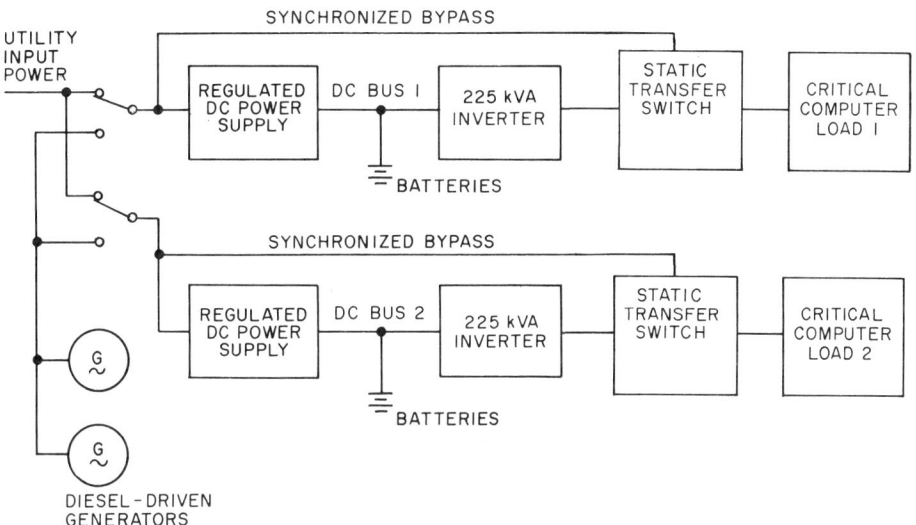

Fig 46
Parallel-Supplied Nonredundant Uninterruptible Power Supply

4.8.8 *Parallel Nonredundant Uninterruptible Power Supply with Static Bypass Switch.* Fig 46 shows a parallel supplied nonredundant uninterruptible power supply system. The installation consists of two discrete systems serving two computers. A synchronized bypass and static switch protect each load in the event of inverter fault. Should voltage be lost to a computer load, the static transfer switch will operate to reestablish voltage in less than one quarter of a cycle, fast enough to be considered continuous power for most loads.

4.8.9 *Combination Static Inverter and Rotating Uninterruptible Power Supply.* A combination static, battery, and rotating uninterruptible power supply system is shown in Fig 47. In normal operation the rectifier is supplied from the prime power source, and the battery is floating on the line. The inverter output frequency is slaved to the prime source and follows it exactly. The solid-state rectifier is supplying direct-current power to the inverter and is also maintaining the batteries at the proper float charge. The generator supplying power to the load can be a standard commercial unit. No flywheels are used in this system.

During a fault in the incoming or normal source of power, the high-capacity battery supplies direct-current power to the inverter. The inverter frequency standard controls the frequency of the synchronous motor within ± 0.1 percent while the battery is maintaining the system. Fig 48 illustrates the stability of the load frequency and voltage under widely varying input conditions. This system shows good voltage and frequency stability under starting loads, even with no base load.

4.8.10 *Combination Static Inverter and Engine Generator Uninterruptible Power Supply.* A typical arrangement of a prime power source, an engine generator source, and a battery source of electric power is shown in Fig 49. During an interruption of normal utility power, the static unin-

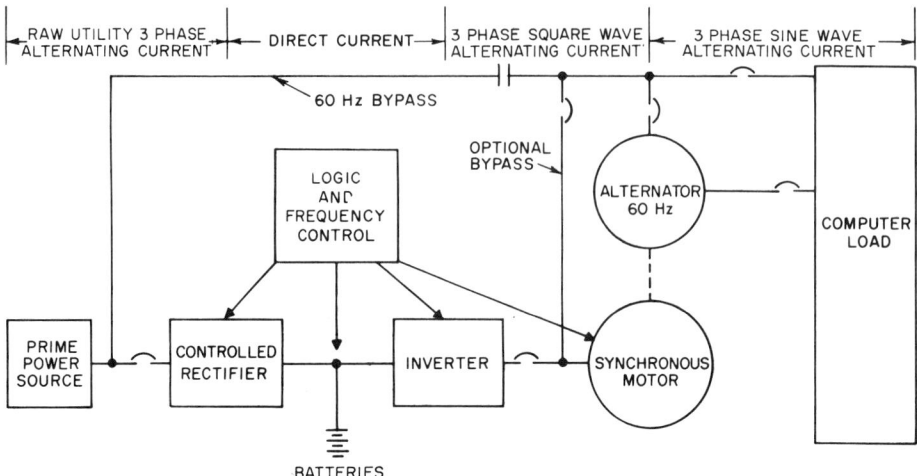

Fig 47
Combination Static, Battery, and Rotating Uninterruptible Power Supply

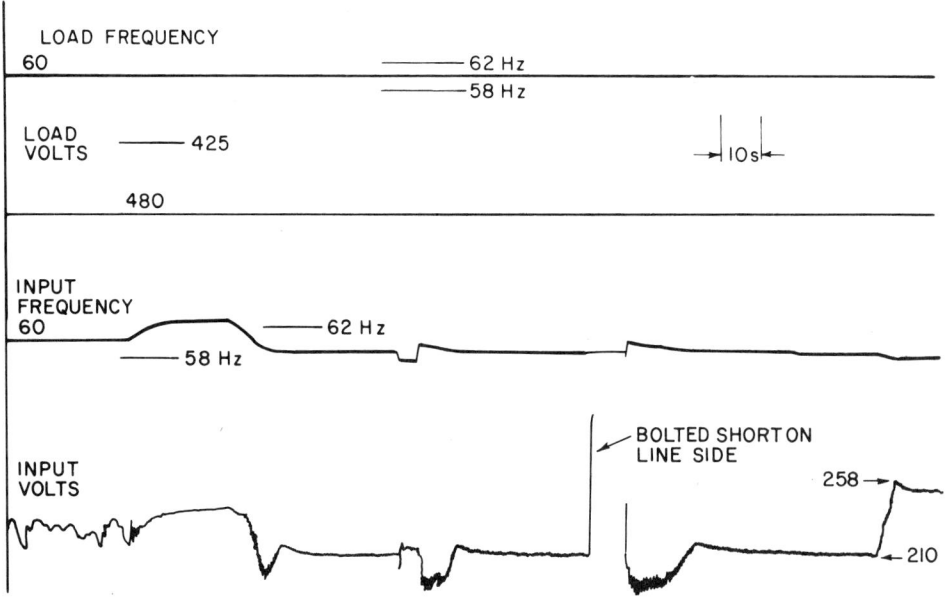

Fig 48
Load Frequency and Voltage Stability of a UPS Under Varying Input Conditions

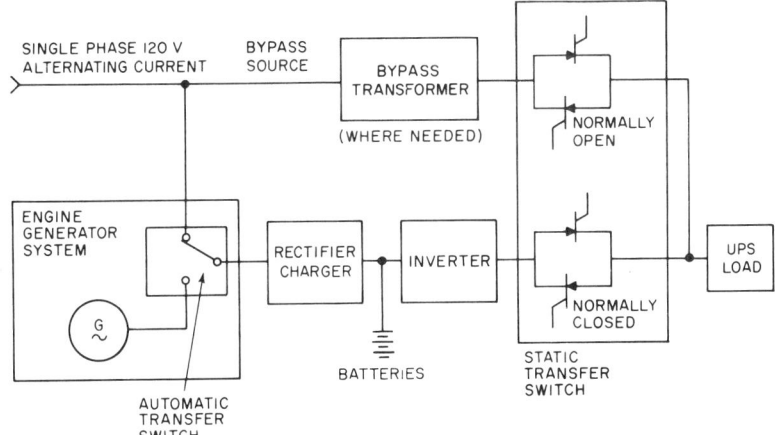

Fig 49
Typical Block Diagram of Combination Uninterruptible
Power Supply and Engine Generator

terruptible power supply provides continuous power to critical selected loads. A time-delay relay prevents the engine generator from starting immediately. After a preset period, the engine generator automatically starts and when the voltage stabilizes, the automatic transfer switch connects the uninterruptible power supply load. The arrangement may be considered because of the possible simultaneous failure of both the electric and gas utilities.

The load is normally served by the inverter initially, but if uninterruptible power supply equipment failure should occur, the load is transferred readily to the bypass source.

To ensure continuity of operation for an extended period, the generator is usually sized 2 to 2.5 times the UPS output rating to account for UPS internal losses, required environmental control systems, and lighting for the critical load. UPS battery charging can be limited while on emergency power to reduce the required generator capacity.

4.8.11 UPS Battery Selection. Inverters require a minimum direct-current voltage for commutation. A commutation failure results in a short circuit which must be cleared by the thyristor fuses. The minimum required direct-current voltage is usually 75 to 80 percent of nominal. A typical inverter nameplate reads 105 to 140 V direct-current input, 120 V alternating-current output. This must be considered when selecting battery sets for uninterruptible power supply systems. It is particularly important for applications where the battery must supply energy for short periods only, because of the relatively large current flowing through the internal resistance of the battery. Uninterruptible power supply systems should have a low direct-current voltage alarm. Low direct-current voltage trip of the inverter should be considered to prevent damage and useless discharging of the batteries.

Economy and reliability will be realized if the volt-ampere switching capabilities of thyristors are optimized

**Fig 50
Typical Battery, Rated 60V, 1800 A · h, at 8 h Discharge Rate**

with appropriate derating for reliability. The primary limitation on thyristors is current, that is, heating. Selection of proper battery voltage can be crucial.

A typical battery installation for emergency power is shown in Fig 50. Such installations may supply direct-current power directly or may be used in conjunction with direct-current motors driving alternators or connected to an inverter for an alternating-current supply.

4.8.12 *Cost to Operate a Nonredundant UPS*. Typical equipment purchase prices for various sized nonredundant UPS systems configured as shown in Fig 42 are provided in Table 20. These prices include a static bypass switch.

It is apparent that as the UPS kVA size increases, the cost per kVA decreases significantly until the load increases to a point (800 kVA, for example) where more than one rectifier/inverter module is required. In general, the cost of a UPS increases in direct proportion to the number of modules required. This direct relationship between number of modules and UPS cost is also demonstrated by an approximate doubling in dollars per kVA for multimodule redundant configurations as compared with single-module nonredundant configurations.

Installation costs for UPS equipment vary greatly from site to site. They can be a major expense when separate battery and rectifier/inverter rooms including ventilation, air conditioning, lighting, emergency shower, and so on are required. Also, the cost of separating the critical and noncritical loads into two distribution systems may be substantial. An installation project cost that is approximately equal to the UPS equipment cost is common.

The future cost of operation and maintenance for a UPS system should not be overlooked in project planning. The yearly electrical energy consumption of a nonredundant 250 kVA UPS, not including the energy used by environmental support systems, currently equals approximately 10 percent of the equipment purchase price. Maintenance costs for the

Table 20
1979 Budgetary Estimates for Uninstalled Nonredundant UPS Systems Exhibiting Typical Performance Parameters of Table 19*

System Output (kVA)	15 min Battery Racks ($)	Inverter/ Rectifier Static Switch ($)	Cost per kVA ($)
10**	4K	12K	1800
25	5K	30K	1400
50	7K	40K	940
100	13K	45K	580
200	26K	57K	420
250	30K	62K	370
400	40K	80K	300
800***	80K	160K	300

* Including all available options except frequency conversion, expandability, EMI suppression, and extended operating temperature capability.
** Single phase.
*** Load is split to be served by two independent 400 kVA nonredundant UPS systems, otherwise a 3-module load sharing redundant type UPS is recommended.

same system will probably add another 10 to 20 percent of the purchase price per year.

4.8.13 *Special Precautions.* A UPS is intended to provide precise and disturbance-free electric power to the critical load. In achieving this end one must not overlook the possibility of "load-caused" power disturbances, particularly when the UPS inverter is serving multiple critical loads.

Load changes due to switching one load "on" or "off" may cause voltage disturbances which are unacceptable to other loads. Similarly, a fault in one item of critical equipment may interfere with or trip off others.

Special attention should be given to this problem when specifying the UPS equipment and the critical load disturbance system. A UPS system can be purchased which will maintain an acceptable voltage during step load changes up to 50 percent. Also, critical loads that may interfere with each other should be physically separated in the distribution system. The installation of isolation transformers or transient suppressors may prevent these undesirable interactions. Cyclic or noncritical loads should not be allowed on the critical bus.

4.9 Standards References

The following standards publications were used as references in preparing this section.

ANSI C84.1-1977, Voltage Ratings for Electric Power Systems and Equipment (60 Hz)

ANSI/IEEE C37.95-1973, Guide for Protective Relaying of Utility-Consumer Interconnections

ANSI/IEEE Std 100-1977, Dictionary of Electrical and Electronics Terms

ANSI/IEEE Std 450-1975, Recommended Practice for Maintenance, Testing, and Replacement of Large Lead Storage Batteries for Generating Stations and Substations

EGSMA GTD2-1971, Glossary of Standard Industry Terminology and Definitions

EGSMA IMFS1-1974, Standards and Recommendations for Installation and Maintenance of Farm Standby Electric Power

EGSMA TDGS1-1972, Standard Specifications for Tractor Driven Generator Sets

EGSMA EGS1-1970, Standard Specifications for Standby Engine Driven Generator Sets

IEEE Std 141-1976, Electric Power Distribution for Industrial Plants

IEEE Std 241-1974, Electric Power Systems in Commercial Buildings

IEEE Std 387-1972, Criteria for Diesel-Generator Units Applied as Standby Power Supplies for Nuclear Power Generating Stations

IEEE Std 485-1978, Recommended Practice for Sizing Large Lead Storage Batteries for Generating Stations and Substations

NECA Electrical Design Library Series 17, Electrical Design Guidelines (1971)

NECA Electrical Design Library Series No 3/74, Emergency and Standby Power Generation (1974)

NFPA No 70-1978, National Electrical Code

NFPA 101-1976, Life Safety Code

4.10 References and Bibliography
4.10.1 References

[1] IEEE Committee Report. Reliability of Electrical Equipment, Pt 1. *IEEE Transactions on Industry Applications*, vol IA-10, Mar/Apr 1974, pp 213–235.

[2] SAWYER, J. W. Gas Turbine Emergency/Standby Power Plants. *Gas Turbine International*, Jan/Feb 1972.

[3] HEISING, C. R., and JOHNSTON, J. F., JR. Reliability Considerations in Systems Applications of Uninterruptible Power Supplies. *IEEE Transactions on Industry Applications*, vol IA-8, Mar/Apr 1972, pp 104–107.

4.10.2 Bibliography

[4] KUSKO, A., and GILMORE, F. E. Concept of a Modular Static Uninterruptible Power System. *Conference Record of the 1967 IEEE Industry and General Applications Group Annual Meeting*, IEEE 34C62, pp 147–153.

[5] LAWSON, L. J. A True No-Break, Off-Line Uninterrupted Power Supply. *Conference Record of the 1967 IEEE Industry and General Applications Group Annual Meeting*, IEEE 34C62, pp 154–158.

[6] GRIFFITH, D.C., and YUEN, M. H. Static No-Break Power for Critical Loads in a Modern Oil Refinery. *Conference Record of the 1967 IEEE Industry and General Applications Group Annual Meeting*. IEEE 34C62, pp 643–652.

[7] KUSKO, A., and GILMORE, F. E. Application of Static Uninterruptible Power Systems to Computer Loads. *Conference Record of the 1969 IEEE Industry and General Applications Group Annual Meeting*, IEEE 69-C5 IGA, pp 635–639.

[8] RELATION, A.E. UPS Systems for Critical Power Supplies. *Conference Record of the 1971 IEEE Industry and General Applications Group Annual Meeting*, IEEE 71C1-IGA, pp 877–884.

[9] WALKER, L. H. Inverter for UPS with Subcycle Fault Clearing Capabilities. *Conference Record of the 1971 IEEE Industry and General Applications Group Annual Meeting*, IEEE 71C1-IGA, pp 361–370.

[10] WOLPERT, T. Uninterruptible Power Supply for Critical AC Loads—A New Approach. *Conference Record of the 1973 IEEE Industry Applications Society Annual Meeting*, IEEE 73CHO763-3IA, pp 595–602.

[11] GROSS, S. Rapid Charging of Lead Acid Batteries. *Conference Record of the 1973 IEEE Industry Applications Society Annual Meeting,* IEEE 73CHO763-3IA, pp 905–912.

[12] HAUCK, T. A. Motor Reclosing and Bus Transfer. *IEEE Transactions on Industry and General Applications,* vol IGA-6, May/Jun 1970, pp 266–271.

[13] HELMICK, C. G. Designing for System Reliability in Large Uninterruptible Power Supplies. *Conference Record of the 1971 IEEE Industry and General Applications Group Annual Meeting,* IEEE 71C1-IGA, pp 371–384.

[14] HELMICK, C. G. Uninterruptible Power Supply Systems—What, Why, Where, and When? Presented at the 34th American Power Conference, Chicago, IL, Apr 18–20, 1972.

[15] KATZAROFF, P. A Base Guide to Uninterruptible Power Systems. *IEEE Conference Record of the 1974 26th Annual Conference of Electrical Engineering Problems in the Rubber and Plastics Industries,* IEEE 74CHO831-8IA, pp 1–6.

[16] KENNY, R. W., McGOVERN, M. J., and TORPEY, P. J. Development of a Gas Turbine-Alternator System for Emergency Power Applications. *IEEE Transactions on Industry and General Applications,* vol IGA-1, Jan/Feb 1965, pp 3–8.

[17] LAWSON, L. J. New Uninterruptible Power System Alternatives Using High Capacity Kinetic Energy Wheels. *Conference Record of the 1973 IEEE Industry Applications Society Annual Meeting,* IEEE 73CHO763-3IA, pp 151–156.

[18] PALKO, E. Standby Generator Specification Chart. *Plant Engineering,* Feb 18, 1971, pp 65–70.

[19] RELATION, A. E. UPS Systems for Critical Power Supplies. *Conference Record of the 1971 IEEE Industry and General Applications Group Annual Meeting,* IEEE 71C1-IGA, pp 877–884.

[20] RELATION, E. A., WINPISINGER, J. L., and MITCHELL, J. T. Uninterruptible Power System Using an Improved Magnetic Voltage Stabilizer. *Conference Record of the 1973 IEEE Industry Applications Society Annual Meeting,* IEEE 73CHO763-3IA, pp 17–23.

[21] RENFREW, R. M. Successful Uninterruptible Power Systems for Computers. *Conference Record of the 1968 IEEE Industry and General Applications Group Annual Meeting,* IEEE 68C27-IGA, pp 787–792.

[22] ROBERTS, A. M. Power Failure Ride-Through for an Inverter System Using Its Own Induction Motor Load as the Energy Source. *Conference Record of the 1968 IEEE Industry and General Applications Group Annual Meeting,* IEEE 68C27-IGA, pp 737–742.

[23] SCHWARM, E. G., and LITTLE, A. D. Computer Uninterruptible Power System with High Speed Static Bypass. Presented at the Summer Power Meeting and International Symposium of High Power Testing of the IEEE Power Engineering Society, Portland, OR, Jul 18–23, 1971.

[24] SUMMERS, G. E. Providing Reliable Power for Computer Systems. *Plant Engineering,* Jan 7, 1971.

[25] SWENSON, E. C. How to Select and Install Standby Electric Plants. *Electrical Construction and Maintenance,* Jan 1963.

[26] The Exciting World of Rechargeable Batteries. *Factory,* Apr 1967, pp 84–87.

[27] System for Orderly Emergency Shutdown. *Modern Manufacturing,* Dec 1969.

[28] Uninterruptible Power System Prevents Computer Downtime. *Rubber World,* Nov 1970, pp 58–60.

[29] The Electric Way to Standby Power. *Plant Operating Management,* Feb 1970, pp 62–65.

[30] Emergency and Standby Power Systems. *Electrical Consultant,* Oct 1971.

[31] The Automatic Transfer Switch Heart of Emergency Power. A Reliability Study of a Power Supply System. The Battery World. *Electrical Consultant,* vol 88, Nov 1972.

[32] Rating Factors for Generating Plants. Tech Bull T-917. ONAN Company, 1400 73rd Avenue NE, Minneapolis, MN 55432.

[33] TERVAY, J. C. Nickel Cadmium Pocket Plate Batteries for Standby Power Applications and Systems. Nife, Inc, 23 Dixon Avenue, Copiague, NY 11726.

[34] Standby Gas Turbine Alternator Package. Publ SD1984. International Harvester Company, 2200 Pacific Highway, San Diego, CA 92112.

[35] Synchronizer. Publ 200-Syn-68 (Gas Turbine). Electric Machinery Manufacturing Company, Minneapolis, MN 55413.

[36] Emergency Lighting Handbook. Radiant Industries, Inc, 10900 Burbank Boulevard, North Hollywood, CA 91601.

[37] GILL, J. D. Transfer of Motor Loads Between Out-of-Phase Sources. *Conference Record of the 1978 IEEE Industry Applications Society Annual Meeting,* pp 1182–1189.

5. Maintenance

5.1 Introduction

Once it has been determined that an emergency or standby power system is justified, available systems must be evaluated to select one which economically satisfies the application requirements. In selecting the proper system, due consideration should also be given to installation requirements and maintainability. After the system is installed and operational, it is imperative that it be properly maintained. This section presents a consolidation of preventive maintenance recommendations.

The goal of preventive maintenance is to maintain equipment in optimum operating condition. The detection and repair of an incipient failure before it develops into a major source of difficulty is one of the primary benefits of a preventive maintenance program. Maintaining accurate records is necessary to set up an optimum schedule of cleaning and inspection. In making up the schedule the following items should be kept in mind:

(1) New equipment must be closely monitored until extensive records indicate whether the schedule can be relaxed or must be tightened.

(2) Old equipment may require more frequent inspection and service.

Infrequently used equipment, such as off line emergency and standby power systems, present a challenge. Preventive maintenance becomes a matter of anticipation rather than detection. In this case the following precautions should be taken:

(1) Make sure that the installation will not be subject to ventilation problems or be obstructed by junk or storage.

(2) Place regular test responsibilities on trained personnel and schedule tests frequently to assure operation when required.

(3) Gasoline and, to a lesser extent, diesel fuels deteriorate when stored for extended periods of time. Inhibitors can be used to reduce the rate of deterioration, but it is sound practice to operate a system utilizing these fuels such that total operating time will result in a complete fuel change cycle every few months.

Although many different emergency and standby systems are on the market today, the general maintenance recommendations presented will provide an indication of preventive maintenance need for various systems. The recommendations are broken down into system components. The components can be reconstructed to form particular systems.

5.2. Internal Combustion Engines

Internal combustion engines commonly available include natural and bottled gas, gasoline, and diesel. Since there are more similarities than differences in maintenance requirements, engines have been lumped together.

Long life and high reliability are characteristics to be expected from these types of prime movers, but only if properly maintained. Preventive maintenance programs will greatly contribute to the service life and reliability.

In establishing a preventive maintenance program for these engines, the best starting point is the manufacturer's service manual. This will provide a guide for specific points to be checked and frequency of inspection. These reference points can then be modified to fit particular installation and operating conditions.

More than any other factor, lubrication determines an engine's useful life. Various parts of the engine may require different lubricants and different frequencies of lubricant application. It is important to follow the manufacturer's recommendations as to type and frequency of lubrication.

Cleanliness should be the foundation of a preventive maintenance program. While there may be minimal wear, there is always a possibility of contamination by corrosive dirt and grit buildup. Dirt is a major cause of equipment failure. Before performing any inspection or service, carefully clean all fittings, caps, filler and level plugs, and their adjacent surfaces to prevent contamination of lubricants and coolants.

Routinely scheduled inspections should include radiator coolant level, antifreeze if utilized, crankcase oil level, fuel supply, and air cleaner. A drain check should be made to eliminate condensed water from fuel tank and filters. The engine should be visually inspected to detect loose nuts, bolts, and other hardware, and leaks at seals, gaskets, and other connections in the fuel, cooling, lubrication, and exhaust systems.

5.2.1 *Typical Maintenance Schedule.* The following maintenance schedule is not intended as a recommendation but is presented as a typical service guide for an 1800 r/min unit.

Every 25 h of operation (or 4 months),

  (1) Adjust fan and alternator belt
  (2) Add oil to oil cup for distributor housing
  (3) Change oil in oil type air filter

Every 50 h of operation (or 6 months),

  (1) Drain and refill crankcase
  (2) Clean crankcase ventilation air cleaner
  (3) Clean dry-type air cleaners
  (4) Check transmission oil
  (5) Check battery
  (6) Clean external engine surface
  (7) Perform 25 h service (above)

Every 100 h of operation (or 8 months),

  (1) Replace oil filter element
  (2) Check crankcase ventilator valve
  (3) Clean crankcase inlet air cleaner
  (4) Clean fuel filter
  (5) Replace dry-type air cleaner
  (6) Perform 25 h and 50 h service (above)

Every 200 h of operation (or 12 months),

  (1) Adjust distributor contact points
  (2) Check spark plugs for fouling and proper gap
  (3) Check timing
  (4) Check carburetor adjustments
  (5) Perform 25 h, 50 h, and 100 h service (above)

Every 500 h of operation (or 24 months),

  (1) Drain and refill transmission

(2) Replace crankcase ventilator valve
(3) Replace one piece type fuel filter
(4) Check valve-tappet clearance
(5) Check crankcase vacuum
(6) Check compression
(7) Perform 25 h, 50 h, 100 h, and 200 h service (above)

5.3 Gas Turbines

5.3.1 *General*. The combustion gas turbine, as with any rotating power equipment, requires a program of scheduled inspection and maintenance to achieve optimum availability and reliability. The combustion gas turbine is a complete, self-contained prime mover. This combustion process requires operation at high temperatures. When inspection marking on stainless steel parts is necessary, a grease pencil should be used. Graphite particles from lead pencils will carburize stainless steel at the high temperature of gas turbine operation.

Starting reliability is of prime concern since a delay in starting usually means the need for the unit has passed.

5.3.2 *Operating Factors Affecting Maintenance*. The factors having the greatest influence on scheduling of preventive maintenance are: type of fuel, starting frequency, environment, and reliability required.

(1) *Fuel*. The effect of the type of fuel on parts is associated with the radiant energy in the combustion process and the ability to atomize the fuel. Natural gas, which does not require atomization, has the lowest level of radiant energy and will produce the longest life of parts. Diesel fuels will provide the next longest life, and the crude oils and residual oils, with higher radiant energy and more difficult atomization, will provide shorter life of parts.

Contaminants in the fuel will also affect the maintenance interval. In liquid fuels dirt results in accelerated wear of pumps, metering elements, and fuel nozzles. Contamination in gas fuel systems can erode or corrode control valves and fuel nozzles. Filters must be inspected and replaced to prevent carrying contaminants through the fuel system. Clean fuels will result in reduced maintenance and extended lives of parts.

(2) *Starting Frequency*. Each stop and start subjects a gas turbine to thermal cycling. This thermal cycling will cause a shortened parts life. Applications requiring frequent starts and stops dictate a shorter maintenance interval.

(3) *Environment*. The condition of inlet air to a combustion gas turbine can have a significant effect on maintenance. Abrasives in the inlet air, such as ash particles, require that careful attention be paid to inlet filtering to minimize the effect of the abrasives. In the case of corrosive atmosphere, careful attention should be paid to inlet air arrangement and the application of correct materials and protective coatings.

(4) *Reliability Required*. The degree of reliability required will affect the scheduling of maintenance. The higher the reliability desired, the more frequent the maintenance required.

5.3.3 *Typical Maintenance Schedule*. The following maintenance schedule is included to show typical maintenance requirements for a combustion gas turbine in an emergency or standby power application:

Weekly,

(1) Check oil level
(2) Check for adequate fuel pressure
(3) Visually inspect all nuts and other fasteners
(4) Check for oil or fuel leakage
(5) Check electrical connections for tightness and corrosion

(6) Check all lines and hoses for external wear

(7) Check air inlet screen for obstructions

(8) Check exhaust system for obstructions

(9) Bleed all drains and collectors to detect obstructions

Every 250 h of operation,

(1) Replace oil filter element
(2) Replace fuel filters as necessary
(3) Check batteries
(4) Blow out air lines and filters with dry low pressure air
(5) Lubricate auxiliary motors

Every 1000 h of operation,

(1) Inspect spark plug
(2) Inspect fuel injectors and combustion parts
(3) Blow low pressure dry air through exhaust and combustor drain lines
(4) Inspect entire engine for unusual discoloration, cracks, wear, or chafing of hoses, lines, and wire, and other unusual operating conditions
(5) Check condition of air inlet filters
(6) Check condition of engine and exhaust thermocouples
(7) Check temperature and speed control units

Manufacturer's service manuals should be used as a starting point for establishing a preventive maintenance program for combustion gas turbines. Adjustments in maintenance intervals can be made from records of operating experience.

5.4 Generators.

Keeping equipment clean is of primary importance in generator preventive maintenance. Dust, oil, moisture, or other substances should not be allowed to accumulate on the equipment. Ventilation ducts should be kept clean to allow maximum cooling air in the generator. The importance of keeping windings clean cannot be overemphasized. Dust, dirt, and other foreign matter can restrict heat dissipation and deteriorate insulation. A layer of dust as thin as 30 mil can raise the operating temperature of generator windings 10°C.

The best method for cleaning a generator of loose and dry particles is to use a vacuum cleaner with proper fittings. Blowing with 30 lb/in^2 compressed air can also be employed, but this method has a tendency to redeposit the particles. Wiping with a soft, clean rag has the disadvantage of not being able to remove dust from grooves and inaccessible places. Buildup of grease and oil can be removed by conservative use of solvents such as carbon tetrachloride or trichloroethane. Neither solvent presents a significant fire hazard; however, trichloroethane is the less toxic of the two. Megohmmeter readings should be taken after cleaning and drying. If the resistance is too low, the cleaning should be repeated.

Regularly scheduled inspections should include checking of all terminations and connections for tightness; checking of all wires for chafed, brittle, or otherwise damaged insulation; checking bearings, brushes, and commutator for proper operating condition. Inspect for creepage of bearing grease inside generator. If moisture has accumulated in the generator, the unit must be <u>dried</u> and strip heaters or other methods must be utilized to prevent the condition from recurring. To dry the unit, external heat should be applied to reduce the moisture content. Internal heat may then be applied by introducing a low voltage current through the windings. The winding temperature must be monitored to prevent insulation damage during the drying operation.

Brushes and connecting shunts should

be inspected for wear and deterioration. Remove the brushes one at a time and check for length. Be sure the brushes move freely in their holders. Brush holders should be checked for proper tension. If the spring tension is strong and is not adjustable, the brush holder should be replaced. Brushes should be replaced when worn down to ½ in. Replace brushes in complete sets, not a single brush. Be sure the shunt leads are properly connected. After placing new brushes in the holders, carefully fit the contact surface of the brushes to the commutator by using first #1, then #00 sandpaper. Do not use emery cloth. Cut the sandpaper into strips slightly wider than one brush. Insert the strip under the brush with the smooth side toward the commutator and draw back and forth around the commutator in the manner of a slipping belt. After the brushes have been seated, clean the carbon dust from the commutator and brush assemblies.

Commutators should be smooth and have a light to medium brown color. A rough or blackened generator commutator may be polished with a commutator dressing stone fitted to the curvature of the commutator. If not available, use #00 sandpaper with a block of wood shaped to fit the curvature of the commutator. Do not use emery cloth. All brushes should be lifted and the generator driven while slowly moving the polishing block back and forth. The mica insulation between commutator bars should be undercut $1/16$ to $1/32$ in. As the commutator wears down, the mica will cause ridges resulting in bouncing of the brushes. If this condition exists the mica must be undercut and the commutator resurfaced by a qualified repairman. Do not use lubricants of any kind on the commutator.

Generator bearings should be subjected to careful inspection at regularly scheduled intervals. The frequency of inspection, including addition or changing of oil or grease, is best determined by a study of the particular operating conditions. Sealed bearings require no maintenance and must be replaced when worn or loose. When the generator is running, listen for unusual noise and feel the bearing housing for vibration or excessive heat.

With proper preventive maintenance a generator will provide reliable and lengthy service.

5.5 Static Uninterruptible Power Supplies

Most uninterruptible power supplies (UPSs) are installed because the load requires an uninterruptible source of power. Therefore, special precautions must be taken when isolating a UPS for maintenance. Be aware of the loads supplied by the UPS and notify necessary personnel that disconnection of the UPS is required for maintenance.

A static UPS is extremely reliable and requires very little maintenance of the inverter and battery charger. See 5.6 for maintenance of batteries.

The UPS should be completely isolated from input and output power, including batteries. Many UPSs may have a unique isolating procedure to prevent an outage to a critical load.

The following items should be included in a regularly scheduled inspection and service schedule:

(1) Be sure all input and output power has been disconnected

(2) Discharge and ground all capacitor terminals in charger and inverter with a grounding stick

(3) Use a vacuum cleaner, and a cloth if necessary, to clean inside of charger and inverter cabinets

(4) Check for liquid contamination

(battery electrolyte, oil from capacitors, etc)

  (5) Tighten all terminations
  (6) Inspect all terminations and control circuits for corrosion
  (7) Check battery condition
  (8) Connect source power and check control circuit power supply voltages per manufacturer's specifications
  (9) Check and adjust voltage output and frequency per manufacturer's specifications

After the UPS has been reconnected, check the output voltage and frequency under load. Simulate a power failure and check for proper system operation.

5.6 Batteries

  5.6.1 All-Liquid Electrolyte Batteries. Batteries should be carefully inspected to reduce contamination. Dust and dirt should be removed from the top of the battery.

The electrolyte level should not be allowed to drop below the top of the plates. Overfilling may cause loss of electrolyte due to expansion and gassing during heavy charging.

Normal charging will not cause excessive loss of electrolytes. Therefore, water alone should be added to correct the level drop due to evaporation and charging. In some places tap water may be used, but in most localities it is best to use distilled water.

During normal use the electrolyte gradually tends to weaken and must be renewed. Do not attempt to increase the specific gravity of weakened electrolyte by adding solution.

  5.6.2 Lead-Acid Batteries. Electrolyte level should not be raised above the bottom of vent tubes. Vent tubes should be kept clear with vent plugs tightly in place. Vent plugs should be cleaned by soaking in warm water.

Every 2 years the battery condition should be determined. Apply an equalizing charge as recommended by the manufacturer. Discharge the battery at the standard ampere-hour rating given in the operating data supplied by the manufacturer. Record individual cell voltages until the majority of the cells reach 1.75 V. Measure the specific gravity immediately. If the specific gravity of the cells are uniform and the battery delivers the discharge current for an acceptable length of time, the battery should be returned to service.

When cleaning batteries, acid buildup should be removed by washing with a mixture of baking soda and water.

  5.6.3 Nickel-Iron-Alkaline Batteries. An alkaline battery that is not used regularly may become sluggish and not be able to deliver its rated capacity. The following steps can be taken to remedy this condition:

  (1) Fully charge the battery
  (2) Discharge the battery through a resistor that can be adjusted to maintain a constant rate until a cell voltage of 0.5 V results
  (3) Short out each cell until the heat of the battery is no more than 5°F above room temperature
  (4) Add water to bring the electrolyte level up to normal
  (5) Charge at normal rate for 16 h
  (6) Discharge at normal rate and record the time it takes to reach 1.0 V per cell
  (7) If the battery reached 1.0 V per cell in less than 5 h, repeat steps 1 through 6

Do not use tools or other equipment to service alkaline batteries if they have been used to service lead-acid batteries.

  5.6.4 Nickel-Cadmium Batteries. The electrolyte from nickel-cadmium batteries is injurious to aluminum, copper, tin, and zinc. Do not allow any cell to stand empty for more than 30 min. Exposure to air will damage plates.

The electrolyte, potassium hydroxide, reacts with carbon dioxide in the air to form potassium carbonate, which, in concentrations of more than a few percent, will decrease the capacity. The following precautions should be taken to reduce the amount of contamination.

(1) Do not open cell vents more often or longer than necessary
(2) Make sure all seals are working properly
(3) Maintain a ⅛ to ¼ in layer of mineral oil on top of electrolyte (do not use mineral oil in batteries of sintered plate construction)
(4) Do not overcharge battery
(5) Keep frequency of electrolyte level adjustment to a minimum
(6) Store electrolyte solution in tightly stoppered containers

The electrolyte must be renewed when potassium carbonate approaches 10 percent concentration.

Tools and equipment used to service lead-acid batteries should not be used on nickel-cadmium batteries. Sulphuric acid will ruin nickel-cadmium batteries.

5.6.5 *Maintenance Interval.* The frequency of inspection and testing of batteries can be affected by the number of charge/discharge cycles and the battery application. A battery used to start an engine may require a shorter inspection interval than a battery utilized in an inverter scheme.

5.7 Automatic Transfer Switches

Automatic transfer switches require maintenance, as do most components of electrical installations. The automatic transfer switch is usually applied where two sources of power are made available for the purpose of maintaining power to a critical load. The necessity of providing safe maintenance and repair of an automatic transfer switch requires shutdown of both power sources or installation of a bypass switch. The bypass switch is used to isolate the automatic transfer switch while maintaining power to the critical load.

5.8 Conclusions

Maintenance is an essential requirement of any electrical installation. The critical nature of emergency and standby power systems applications dictate the importance of properly maintaining the equipment involved. A good preventive maintenance program will aid in maintaining a reliable system.

5.9 References

A version of this section appears in [1].

[1] McWILLIAMS, D. W. Maintenance of Emergency and Standby Power Systems. *Conference Record of the 1976 Industrial and Commercial Power Systems Technical Conference,* IEEE 76CH1081-91A, pp 16–20.

6. Protection

6.1 Introduction

This section is intended to discuss recommended practices and guidelines in applying protection to emergency and standby power systems. Standard practice for protection of devices should always be given full consideration when applying devices for emergency and standby use. However, for this type of application, reliability should often be given extra consideration concerning critical loads and the power supply to the loads. All national, state, and local codes and standards applicable to protection of the components making up the emergency or standby power system must be adhered to. It is not the intent of this section to list all applicable codes and standards, but some pertinent ones are mentioned where necessary.

Protection for individual components that make up the most common emergency and standby power systems is discussed, with emphasis on maintaining the required integrity and reliability of the system. Proper application of systems as a whole is discussed in other sections and should be considered an important part of system protection.

6.2 Short-Circuit Current Considerations

There are many areas of concern for protection of components that make up an emergency and standby power system, but the one given the most attention is that of short-circuit current. Studies should be made to determine the current magnitudes available throughout the emergency and standby power system, especially at switching and current interrupting devices.

In most cases, emergency and standby power systems do not have as much fault current available as do normal power systems. When this is true, the magnitude of the short-circuit current available from the normal system determines the required interrupting or withstand rating, or both, of the system components. An emergency or standby generator should be evaluated as to whether or not it will supply enough fault current to open a branch circuit overcurrent device that is coordinated with a normal power source feeder overcurrent device. Subtransient current may be very important in this case. Fig 51 shows a simple illustration of the form the rate of decay of fault current

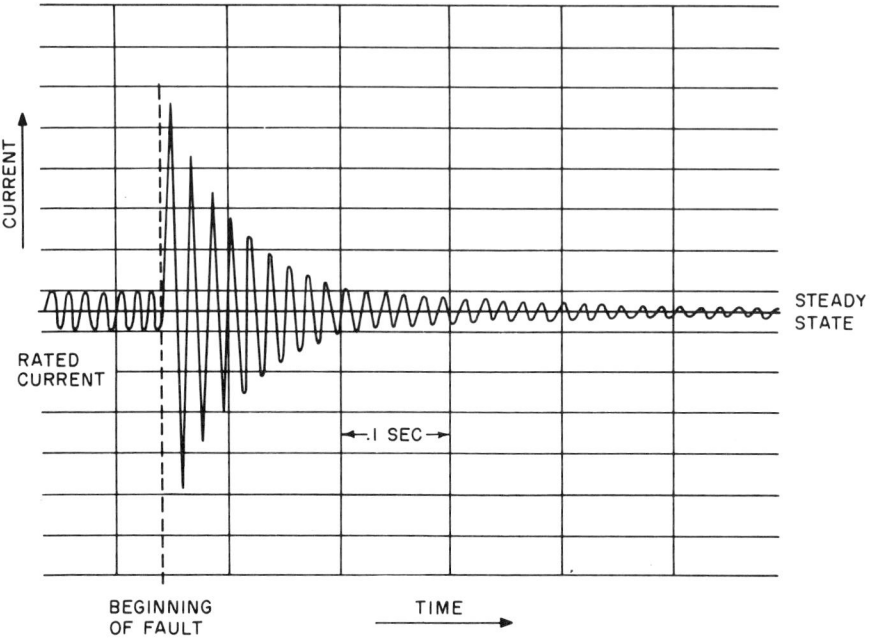

**Fig 51
Fault Current Delay**

from a limited source might take. The rate of decay will determine the kind of coordination necessary between main and branch circuit overcurrent devices and the types of devices to use, such as circuit breakers or fuses, to achieve proper coordination and selectivity. Generator short-circuit characteristics should be obtained from the manufacturer.

6.3 Transfer Devices

6.3.1 *Codes and Standards.* Some applicable codes and standards are listed as follows:

ANSI C33.122-1976, Standard for Automatic Switches (UL 1008)

CSA Standard C22.2 No 178 (Canadian Standard), Automatic Transfer Switches

NEMA Industrial Controls and Systems, Part ICS-2-447, Automatic Transfer Switches

NFPA No 70-1978, The National Electrical Code

NFPA 76A, 1977, Essential Electrical Systems for Health Care Facilities

Codes and standards applicable to protection of transfer devices vary somewhat between different usages of the equipment. The National Electrical Code is vague in requirements for transfer equipment in emergency and general standby power generation systems. Specifically, NEC Articles 750-8 and 517-66(f) deal with overcurrent protection of transfer switches. Underwriter Laboratories Standard 1008 pertains to construction and testing of automatic transfer switches. As of 1974, UL 1008 became effective and only those automatic transfer switches that can comply with the required load duty and fault current withstand ratings are eligible for listing by Underwriter Laboratories.

6.3.2 *Withstand Ratings.* Transfer switches must be given special consideration over normal branch circuit devices on application. The design, normal duty, and fault current ratings of the transfer switch play an important part in its application and protection scheme. It must be capable of closing into high inrush currents, withstanding fault currents without damage or contact separation, and be capable of severe duty cycle in switching normal rated load. All are application considerations, but emphasis will be mainly on fault withstandability. The coordination of overcurrent protection devices to match transfer switch ratings under fault conditions is one of the most important application aspects in providing reliable operation of a standby or emergency power system.

The destructive effects of high fault currents consist of two components: (1) magnetic stresses that attempt to pry open the switch contact, and (2) heat energy developed which can melt, deform, or otherwise damage the switch. Either or both of these components can cause switch failure.

A fault may cause a substantial voltage drop which can be sensed by the relays in the automatic transfer switch. It is imperative that the switch contacts remain closed until protective devices can clear the fault. Separation of the contacts, prior to protective device operation, can develop enough arcing and heat to damage the switch. Time delay to prevent immediate transfer of mechanically held mechanisms, and contact structures specifically designed to utilize electromagnetic forces to increase contact pressure, combine to provide reliability necessary in automatic transfer switch operation. It follows, of course, that proper application of the switch within its withstand rating is important to prevent contacts from welding together and any other circuit path joints and connections from overheating or deforming, thus prolonging the life and increasing reliability of the switch.

UL 1008 has tables which list minimum short-circuit currents that transfer switches must withstand and optional short-circuit currents which may be assigned. Table 21 shows typical manufacturers' withstand current ratings when protected by current limiting fuses or circuit breakers.

Care must be taken to account for asymmetrical and instantaneous peak fault currents when applying transfer switches. The user must not be misled by specifications of seemingly high ratings that may actually be asymmetrical or peak current ratings. Symmetrical rms amperes should be used when coordinating a protective device with the time-current characteristic.

Manufacturers should be consulted in determining the method of testing applied to transfer switches. Whether a fuse or circuit breaker was used and the X/R ratio are important aids in determining the best protection in actual application. Current limiting fuses, for example, when applied, would considerably limit the duration of short-circuit current compared with application of circuit breakers.

6.3.3 *Significance of X/R Ratio.* It is the X/R ratio of a circuit that determines the maximum available peak current and thus the magnetic stresses that can occur. As the X/R ratio increases, both the fault withstandability of the switch and the capability of an overcurrent protective device become more critical. The power factor, X/R ratios, and their relationships to peak current can be found in Table 2 of ANSI/IEEE C37.26-1972 (R 1977). Many

Table 21
Example Withstand Current Ratings for Automatic Transfer Switches

Switch Ampere Rating	Available Symmetrical Amperes RMS at 480 V AC and X/R Ratio of 6.6 or Less			
	When Used with Class J and L Current Limiting Fuses		When Used with Molded Case Breakers	
	WCR	Maximum Fuse Size (A)	WCR	Maximum Breaker Size (A)
50	24 000	125	5 000	125
100	100 000	300	22 000	600
225	100 000	600	22 000	600
400	100 000	800	35 000	1200
600	200 000	1200	42 000	2500
800	200 000	1200	42 000	2500
1000	200 000	2000	55 000	2500
1200	200 000	2000	55 000	2500
1600	200 000	3000	85 000	2500
2000	200 000	3000	85 000	2500

Note: X/R ratio, size of overcurrent protective devices, and withstand current ratings vary depending upon the manufacturer.

times a circuit breaker symmetrical current interrupting rating or a transfer switch withstand rating must be reduced if applied at an X/R ratio greater than what the device safely withstood at test.

When current limiting fuses are employed as protective devices for switches, the peak instantaneous let-through current passed by the fuse must be equal to or less than the instantaneous peak rating of the switch. Also, the fuse interrupting rating and test X/R ratio should be greater than the circuit available fault current and rated X/R ratio, respectively. Tables 22 through 24 show how X/R ratios vary for circuit breaker interrupting ratings, fuse interrupting ratings, and transfer switch withstand ratings.

6.3.4 *Transfer Switch Dielectric Strength.* An area of concern not to be overlooked is the dielectric strength of the switch to be applied. UL testing requirements are designed mainly for safety, but again reliability must be considered. After withstanding a fault, a transfer switch complying fully with UL 1008 test requirements could subsequently fail due to the transient voltage surges if proper surge protection is not employed.

For the user to intelligently and correctly apply a transfer switch within its withstand rating, he should obtain from the manufacturer full information about the test circuit, short-circuit magnitudes in rms symmetrical amperes, circuit X/R ratio, and voltage on which the withstand rating is based.

6.3.5. *Protection with Circuit Breakers.* Coordination of main and branch circuit devices should be given special consideration in emergency and standby power systems if loads are critical. In addition to transfer switch protection, reliability is maintained by proper selectivity in breaker tripping.

Normally, some time delay is required in the main (service) breaker to achieve selectivity. This may require application

Table 22
Molded-Case and Power Circuit Breaker Interrupting Requirements Per UL Std 489 and IEEE Std 20-1973

Interrupting Rating (Symmetrical)		Test Power Factor (%)	X/R Ratio
Molded-Case Breakers	Power Breakers		
10 000 and less	—	45–50	1.73–1.98
10 001–20 000	—	25–30	3.18–3.87
20 001 and more	—	15–20	4.9–6.6
—	All ratings	15 max	6.6 min

Table 23
Automatic Transfer Switch Withstand Requirements Per UL Std 1008

Withstand Test Available Current (Symmetrical Amperes)	Test Power Factor	X/R Ratio
10 000 or less	40–50	1.73–2.29
10 000 to 20 000	25–30	3.18–3.87
20 001 and more	20 maximum	4 minimum

Table 24
Fuse Interrupting Test Requirements Per UL Stds 198B, 198.2, 198.3, 198.4, 198H, and ANSI C97.1-1972

Fuse Class	Interrupting Test Current (Symmetrical)	Test Power Factor	X/R Ratio
H*	10 000	45–50	1.73–1.98
K	50 000 100 000 200 000	20 maximum	4.9 minimum
J	200 000	20 maximum	4.9 minimum
L	200 000	20 maximum	4.9 minimum
R	200 000	20 maximum	4.9 minimum
T	200 000	20 maximum	4.9 minimum

*When rated above 100 A.

of a power circuit breaker or a special molded case breaker with an adjustable short time delay trip feature. Obviously, proper coordination will prevent nuisance starting of an emergency generator on tripping of the breaker from the main power source. The added cost of the main breaker is usually justified by the increased degree of service reliability. Figs 52 and 53 depict a simple illustration of this arrangement. Breakers A and B would be provided with some time delay in the short trip region, while breaker C would be provided with an instantaneous trip to achieve proper coordination. Similarly, the nonessential load feeder breakers would be provided with instantaneous trips to coordinate with the main breaker.

The transfer switch protected with a circuit breaker may have to carry fault current for a relatively long period of time. The fault current magnitude, because of the switch I^2t rating, must be relatively low to prevent exceeding this thermal rating. For instance, a transfer switch may have a rating of 100 000 A symmetrical when used with a specific current limiting fuse, but may only have a withstand rating of 30 000 A symmetrical when used with a specific circuit breaker with an instantaneous trip. The same reasoning would apply for breakers with different fault clearing times, especially considering short time delay trips versus instantaneous trips. ANSI C37.16-1973 recognizes this by requiring reduced short-circuit current interrupting ratings for short time delay applications. For example, a 225 A frame low voltage power circuit breaker, with an instantaneous trip unit, has a 22 000 A symmetrical interrupting rating at 480 V. The same breaker, equipped with a short time delay trip, has a reduced rating of 14 000 A symmetrical at 480 V.

Planning for future expansion can save

IEEE
Std 446-1980

EMERGENCY AND STANDBY POWER SYSTEMS

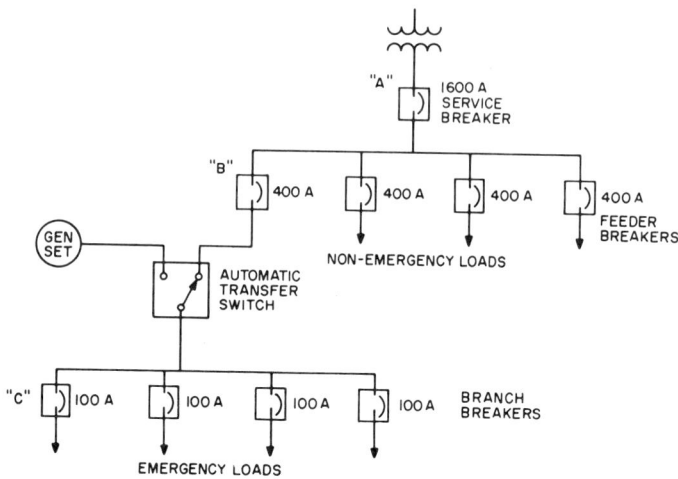

**Fig 52
One-Line Diagram for Fig 53**

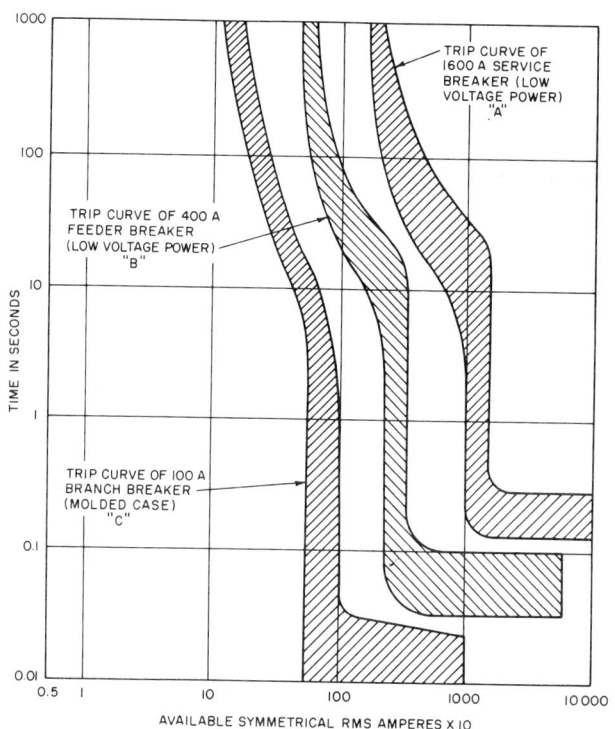

**Fig 53
Coordination of Protective Devices in Fig 52**

problems at a later date, especially when short-circuit current availability increases demand, such as might be the case when an electric utility expands its system. One solution, when the problem of short-circuit current availability exceeding transfer switch or circuit breaker rating occurs, is to apply current limiting fuses to the existing breaker. This offers an economical compromise of limiting short-circuit current and I^2t let-through and maintaining some operating flexibility.

6.3.6 *Protection with Fuses.* Fuses can safely interrupt higher short-circuit currents faster than circuit breakers. An advantage of breakers over fuses is gang operation of poles eliminating the possibility of single-phase conditions. Fuse peak let-through current and I^2t energy let-through should be coordinated with the same characteristics of the transfer switch applied. These characteristics vary among fuse manufacturers and should, therefore, be known for each particular fuse used. A transfer switch may be sized and rated for operation in series with a specific fuse. If another class of fuse, with the same ampere rating and interrupting rating, is substituted, the transfer switch may fail under fault conditions. The substitute fuse may have permitted a higher I^2t energy or peak current let-through. When current-limiting fuses can be used with automatic transfer switches, the switch rating data, in addition to the test circuit data, should include the maximum instantaneous peak let-through current and I^2t energy let-through associated with the recommended fuse class. The designer can then coordinate the transfer switch ratings with any manufacturer's fuses. He can also choose a different fuse class with time-current characteristics suitable for a particular coordination problem and still keep let-throughs within the transfer switch capabilities.

Simple example illustrations of fuse

**Fig 54
Emergency Power Supply with All Fuse Protection**

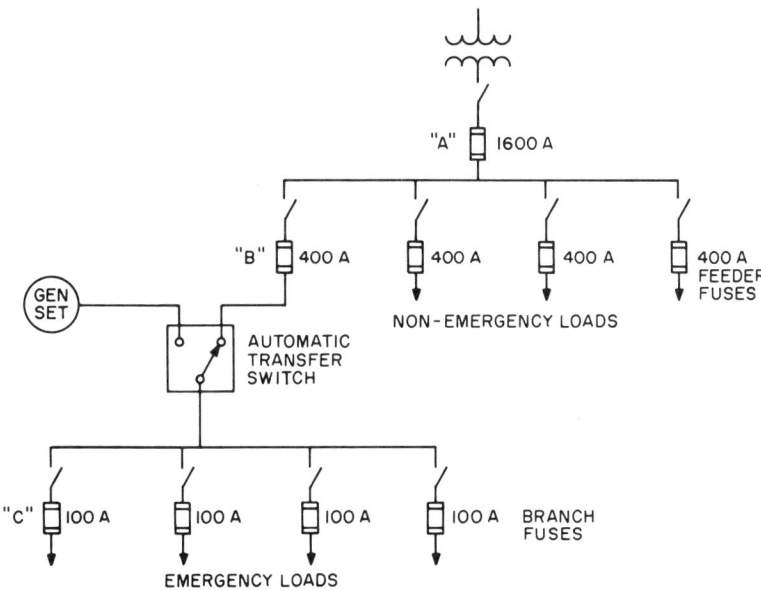

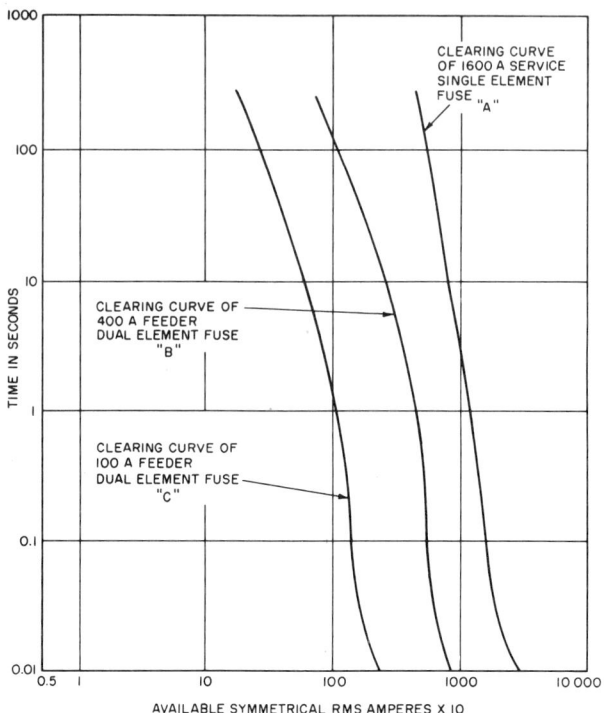

**Fig 55
Coordination Chart of Emergency Power System with All Fuse Protection**

protection and coordination are shown in Figs 54 and 55. As indicated in Fig 55, fuses A, B, and C coordinate in the long time operating range from 0.01 s (approximately ¾ cycle) to 1000 s. To check fuse selectivity in the short time or current limiting range (operating time less than ½ cycle), selectivity tables or I^2t let-through curves must be consulted.

6.3.7 *Static Transfer Switches.* Available short-circuit current is especially critical in application of static transfer switches. Fuse clearing time and switch rating must be properly coordinated to minimize the effect of a fault on the load as well as the switch. Fast opening of the switch (removal of gating voltage) and fast fuse operation are important in maintaining reliability. Care must be taken to coordinate fuse ratings with the different applied power sources and load equipment. Circuit breakers in general are often considered too slow compared with fuses for proper protection and operation of static equipment. Another consideration is that fast fuse operation may prevent an inverter from going into current limiting.

As with all static devices, transient voltage protection should not be overlooked. The thyristors of a static transfer switch are vulnerable to this hazard if transient suppression is not applied. It is well to note here that UPS systems, with such devices as battery chargers and inverters, should be given special consideration in protection from voltage transients. Equipment manufacturers can

PROTECTION

often provide valuable assistance in this area when application requirements are known.

6.4 Generator Protection

When an emergency generator is running, it is the most critical and vital element in a power supply. The protection scheme chosen must ensure reliability as well as protection, such that protection sometimes takes on an added dimension. A user must rely on operating experience and sound judgment in determining whether or not standard protection schemes allow the required reliability. For instance, whether or not an installation is remote or locally operated may determine if an alarm will suffice where normally an automatic trip is employed. Ground faults occurring in any part of an emergency system can cause problems in coordinating generator overcurrent devices with those of feeder circuits. The evaluation under worst case conditions of consequences and cost of loss of critical load versus destruction of the emergency power source may be required. Caution, however, must be exercised when emphasizing reliability to ensure that codes and standards are not overlooked.

6.4.1 *Codes and Standards.* The National Electrical Code, regarding emergency or standby generators, is very general and brief in its requirements on protection. Manufacturers generally provide the basic protection needs of the equipment requiring the user to specify certain options. IEEE Std 242-1975 (IEEE Buff Book) provides overall recommended practices for generator protection, but the user must still consider applications associated with emergency and standby use of the generator.

6.4.2 *Main Winding Protection.* Main breakers are typically standard equipment with emergency or standby generators. Where only one transfer switch is to be supplied by the generator, it may be desirable to omit the circuit breaker depending on the degree of reliability required. When a main breaker is to be used, a power breaker offers easier coordination with its adjustable tripping characteristics in the long, short, and instantaneous ranges. However, molded case breakers are more economical and smaller in size for a given rating but more difficult to coordinate with other overcurrent devices. Molded case breakers are available on the market today with solid-state and mechanical time delay unlatching, but at higher costs. Fuses, of course, are the simplest and most economic overcurrent devices available today. They are also highly reliable when applied properly, but do not offer the flexibility of a circuit breaker. An obvious compromise to the above would be a circuit breaker with integral current limiting fuses, when a breaker is to be applied where available fault current can exceed its interrupting rating.

On small generator applications, selectivity in breaker operation can be a problem when only limited short-circuit current is available, as mentioned in 6.2. Consideration should be given to utilizing a regulator to allow sustained fault current to gain the required selectivity. Collapse of a generator due to slow operation of an overcurrent tripping device would deenergize all critical load and negate the inherent reliability of a properly coordinated system. The rate of generator current decay might take on the form shown in Fig 51.

With large machines, protection can become very sophisticated depending upon investment. Normal protection schemes for generators can vary from a single, simple, molded case circuit breaker to a larger power circuit breaker with complex relaying for trip initiation. IEEE Std 242-1975 (Buff Book) outlines standard protection schemes for generators.

The neutral grounding scheme employed warrants special attention in the application of emergency generators. This is an area where many arguments favor ungrounded or high resistance grounded systems to gain increased reliability. When the engine generator neutral is grounded, problems can arise when multiple neutral-to-ground connections exist, such as incomplete sensing of ground currents and nuisance tripping. Section 7 explores in detail various aspects of grounding and handling of the system neutral.

Surge protection should not be overlooked, especially where vacuum breakers are used. The larger and higher voltage rating generator used, the more critical this protection becomes. It is less critical with smaller units with lower voltage ratings (600 V and below) since more insulating capacity is normally built in.

As mentioned in 6.2. an emergency or standby generator may not have as much available fault current as the normal power supply. Another problem often overlooked in this area relates to the higher impedance of the emergency or standby unit. Besides the problem of availability of fault current, other complications which may arise are the following: (1) switching a capacitor along with critical loads may overexcite the generator; (2) waveform distortion caused by static power converters can cause incorrect operation of the generator voltage regulator, leading to instability in the system; and (3) high current harmonics associated with converters can cause overheating of the generator.

6.4.3 *Rotor and Excitation System Protection.* A field circuit breaker is a positive means of protecting the rotor and excitation system from damaging overcurrents due to generator misapplication or component failure within the excitation system. The user should not assume that a field circuit breaker provides adequate protection for the phase windings. This breaker should in no way replace a main breaker in the generator output.

6.4.4 *Parallel Operation.* Protection of generators when two or more are operating in parallel is necessarily more standardized than that of only one generator. Some essential protection considerations are reverse power flow, synchronizing check, and load shedding.

Application of reverse power relaying must be carefully handled since sensitivity settings of relays can cause nuisance trips or generator damage. If a generator is to parallel with an electric utility, for example, fast tripping may be necessary for reverse power flow into the utility system to prevent generator overload. This subject is discussed also in 6.5.1.

Load shedding, of course, speaks for itself in the prevention of overload. This subject is also treated more thoroughly in 6.5.1.

6.5 Prime Mover Protection

6.5.1 *General Requirements.* The most direct form of overload protection, still maintaining some degree of reliability, would be load shedding. Depending on the severity of stability problems, breaker position or frequency sensing might be used to initiate action. Instantaneous automatic load shedding where multiple generator sets are used, when one or more generators are lost, would assure available power for the remaining, more critical loads. On smaller systems, especially where only one generator is used, frequency sensing to shed load might provide a more reliable power supply if upset conditions do not always require load shedding. A combination of the two methods would allow a system to instantaneously shed selected loads, with frequency sensing employed as back up to shed additional load as necessary. It is common practice in some cases to employ underfrequency relays as secondary sensing devices to trip selected load breakers in multiple steps, with a time delay between each step. A stability study would determine frequency settings to be sensed at each successive step. The study would also determine how fast load shedding should occur and thus determine the type and speed of equipment to be used. The scheme would normally be designed to maintain enough generation to prevent total blackout regardless of the load conditions.

Overload protection for a prime mover can also be aided by assuring that frequency sensitive voltage regulation is used. Maintaining a constant voltage-frequency ratio minimizes effects of overload and allows a unit to more easily regain normal voltage and frequency after an overload. The output voltage of the generating set would decrease in proportion to the frequency (prime mover speed). The use of a nonfrequency sensitive voltage reference could require, in some cases, a load reduction from 50 to 60 percent to allow return to rated speed. Although some manufacturers can provide either type of voltage reference, it may be up to the user to specify his preference. Application of frequency sensitive voltage regulation, however, should not overshadow the importance of proper matching of generator or prime mover torque characteristics.

Reverse power relaying is an important form of protection for prime movers. It will prevent motoring when generator sets are operating in parallel and in another application prevents overload of the generator set by fast relaying when power flow is into an electric utility system. In prevention of generator motoring, the user must be aware that some prime movers are less susceptible to damage than others. Sensitivity is more critical, for instance, on turbines than reciprocating engines and nuisance tripping can occur if this is not accounted for.

Protection against motoring of a prime mover guards against overheating or cavitation of blades on a turbine and possible fire or explosion from unburned fuel in a reciprocating engine. A relay to detect reverse power flow would normally be applied as backup protection to mechanical devices designed to detect these conditions. A time delay can be applied to prevent nuisance tripping on momentary reverse power surges such as may occur during synchronizing. Some typical values are listed below showing reverse power required to motor a generator when a prime mover is being spun at synchronous speed and at no input power:

Condensing turbine	3 percent of nameplate, in kilowatts
Noncondensing turbine	≥3 percent of nameplate, in kilowatts
Diesel engine	25 percent of nameplate, in kilowatts
Hydraulic turbine	0.2 to 2.0 percent of nameplate, in kilowatts

6.5.2 *Equipment Malfunction Protection.* Standard protective devices and numerous options are provided with prime movers by manufacturers. The equipment investment and nature of critical loads determine how to apply this protection. Some shutdown devices might be considered for alarm only, such as where installations are attended, and a malfunction can be quickly investigated. When it is determined that a shutdown is necessary for a malfunction condition, protection integrity can be maintained and power supply reliability enhanced if the malfunction is such that an alarm and subsequent shutdown can be employed.

High water temperature, high oil temperature, low oil pressure, overspeed, high exhaust temperature, and high vibration are typical examples of malfunctions that lend themselves to two-level protection as suggested above. On large machines, this is usually an insignificant investment to make. It is common practice in many installations and often required by machine manufacturers to also provide meters for continuous monitoring of the above parameters by personnel on site or at remote locations. When remote-controlled machines are located in areas not easily accessible, readouts of individual machine parameters become more significant. Cases have occurred where engines and turbines have been damaged because operating personnel with remote control capability restarted the units or continued operation of the units under malfunction conditions without knowing specifically which malfunctions existed.

Large generator sets requiring sophisticated and complex control systems often utilize UPS systems for control power. Protection of power supplies of this nature is vital to the total machine protection.

Reciprocating engines and turbines require different philosophies in protection by nature of their design and operation. Smaller reciprocating engines typically have protection furnished for high water temperature, low oil pressure, overspeed, and failure to start. Larger units might also include high oil temperature, high vibration, antimotoring, and protection for sophisticated control and excitation systems.

Protection for combustion turbines would add such protective devices as failure to light, failure to reach self-sustaining speed, and exhaust temperature limits and control. Overspeed and vibration are obviously more critical on turbines due to high speed operation.

Electric-motor-driven auxiliary equipment on large generator sets, such as lube oil pumps and cooling water pumps, are vital to the protection and reliability of the generator set. Reliability might be increased and protection maintained in locally operated installations by removing thermal overload trips from motor control circuits and alarming only on an auxiliary motor overload. In this instance, lube oil temperature and levels, cooling water temperature, and auxiliary motor loading should be closely monitored to protect equipment investment.

6.5.3 *Fuel System Protection.* The importance of protection for fuel systems needs very little explanation regarding the reliability of the power supply. For instance, it is obvious that low pressure and level alarms can prevent needless shutdowns or failures of emergency generator sets requiring a high degree of reliability. Building codes and fire insurance regulations aid in determining optimum locations of self-contained fuel systems. Reference is made to Section 4 for storage recommendations, and NFPA No 30-1973, Flammable Combustible Liquids Code, gives some requirements on protection for fuel piping.

It is well to note that gasoline and diesel fuels can deteriorate if left standing unused for long periods of time. Some provision should be made for periodically burning or replacing the fuel at regular intervals.

6.6 Electric Utility Power Supply

Protection schemes vary for various types of electric utility supply connections, but the usual objectives are to protect the main utility supply from the adverse effects of faults between the utility main disconnect device and the service entrance equipment and the adverse effects of faults in the utility's system.

If the electric utility supply is in parallel or floating with the industrial system where in-plant generation is applied, reverse power and directional overcurrent relaying are important to prevent power flow or fault current from feeding into the utility system. Many utilities will not allow parallel operation with industrial power supplies and in most cases economics prohibit using an electric utility supply as "standby" only. The most common application of an electric utility supply for standby use is that of dual or multiple services from the utility where any one service is capable of serving the entire load. Protection applied by the industrial user will depend largely on the ownership of service entrance or substation equipment, or both.

Ground fault sensing can sometimes be a problem if not carefully planned. Reference is made here to Section 7 for special sensing methods.

6.7 Uninterruptible Power Supply (UPS)

6.7.1 *Battery Protection.* Batteries supply the reliability inherent in an uninterruptible power supply and protection should be given prime consideration. Some important areas of concern are overcharging, discharge rate and limits, ambient temperature, and ground detection.

Overcharging causes gassing and degradation of lead-acid batteries. Fumes can also cause corrosion of terminals. Sustained high temperature operation of lead-acid batteries causes internal corrosion of grids and plates. A low voltage alarm and trip, of course, can prevent needless discharging of batteries and consequent damage. Alkaline batteries such as nickel-cadmium are generally accepted as durable and rugged, both chemically and mechanically. Vented-type cells, although creating very little gas in normal operation, do require periodic additions of electrolyte, even if infrequently. Some maintenance is required.

Prolonged overcharge currents can cause substantial gassing of vented cells and heating of sealed cells. Sealed cells sometimes incorporate devices to permit rapid charging without excessive pressure buildup or overheating. Overcharging is protected against in some cells by automatically shunting charging current.

Ambient temperature increases directly affect some lead-acid batteries by increasing water consumption, grid corrosion, and hydrogen evolution. This is to say that operation in higher than recommended ambients shortens battery life.

Maintenance for protection of batteries during storage can vary depending on type, such as requirements for periodic charging and discharging and additions of electrolyte. For instance, nickel-cadmium batteries, even if vented, require very little attention if filler caps and valves are kept closed. If water is added to a battery before it is placed in storage where it may be subject to low temperatures, care should be taken to assure the water is thoroughly mixed with the electrolyte to prevent freezing. Because nickel-cadmium batteries can be stored for long periods of time in a discharged state without detrimental effects, it is the preferred method to avoid damage due to accidental shorting of the terminals. Generally, all batteries should be stored in level and up-

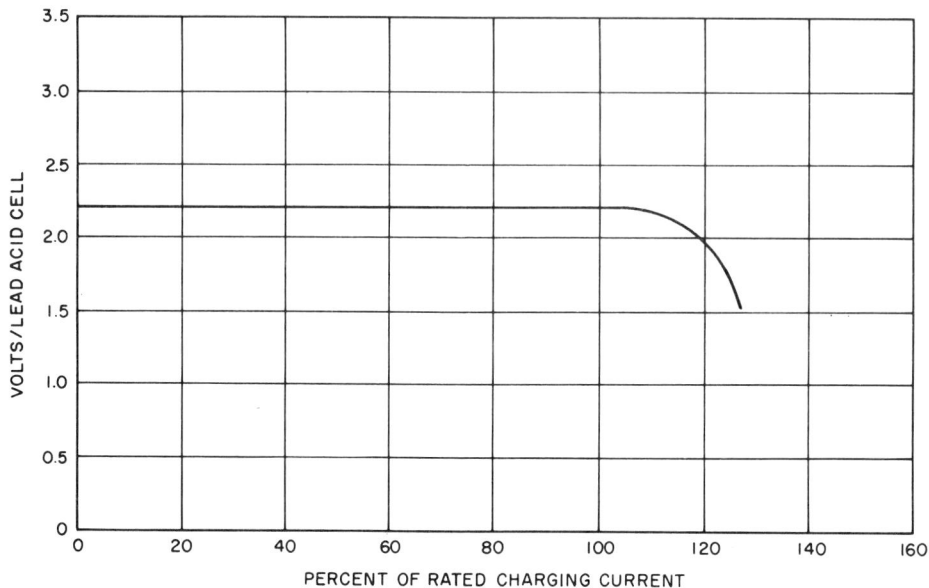

Fig 56
Battery Charger Regulation Curve

right positions in clean, cool, dry locations.

6.7.2 *Battery Charger Protection.* Various protective features for battery chargers include current limiting output, surge suppression, and fuses and circuit breakers. A current limiting output provides overload protection for the charger. A typical limit might be 125 percent of the charger rating and for short periods of time the battery can supply the required excess output. Fig 56 shows a typical regulation curve. Some current limiting features provide automatic charger shutdown when operating into a short circuit or allow the charger to operate into a short circuit without tripping breakers or opening fuses in the input or output circuits. Surge suppression can be provided by the manufacturer on the charger input and output to guard against line transients. Input and output circuit breakers add to overload protection as well as provide flexibility.

Optional devices on battery chargers might include ground detector switch or voltmeter, ground detector lights, power failure disconnect or alarm relay, and high and low dc voltage relays. An ac power failure disconnect relay actually protects the battery from unnecessary discharge back through the charger.

6.7.3 *Inverter Protection.* Inverters are commonly protected on input and output with circuit breakers or fuses. Prolonged short-circuit conditions, out of phase switching, and accidental reverse polarity connections are examples of conditions protected for by overcurrent devices like breakers and fuses. Current limiting on output circuitry, as with battery chargers, is provided by most manufacturers.

Inverters can be supplied with some overload capability built in. Typical capabilities might be 125 percent overload for 10 min and 150 percent for 10 cycles. Undervoltage sensing can be provided to shut down an inverter if battery voltage drops below a predetermined value. When nonlinear loads, such as motors drawing peak currents, can occur, it may be more economical to transfer the load to another source than to increase the inverter rating.

An area of protection often overlooked in application of inverters is adequate ventilation. Under normal operation, inverters can give off a considerable amount of heat. Care should be taken not only to provide adequate ventilation, but also to prevent needless blocking of this ventilation.

6.8 Equipment Physical Protection

Engine-generator sets and fuel supplies and their associated equipment require careful planning in application to prevent damage from physical abuse and environmental conditions. Radiator cooling on engines should be given special consideration, when possible, since a self-contained cooling system does not require external piping connections that could be subject to damage.

Most other devices and equipment discussed above, applied in emergency and standby use of equipment, are usually applied and located where only qualified personnel have access, such as chargers, inverters, etc. When this is not possible, enclosures should have provisions for locking. Manufacturers must be thoroughly knowledgeable of the location and environmental conditions for application of proper enclosures and seismic protection. For example, corrosive or humid atmospheres may require heaters or special precautions to protect generator windings. Location in proximity to large reciprocating machinery might require special seismic protection. Fuel systems, of course, require special consideration in protection from leaks, contaminants, etc.

Manufacturers can usually provide suitable racks for batteries that provide protection and easy maintenance. It is important to have properly sized racks that space cells to minimize overheating and corrosion problems, enhance easy maintenance, etc. Manufacturers can usually assist in designing installations for seismic protection, such as where earthquakes can be a problem.

6.9 Grounding

Reliable ground-fault protection schemes require a careful analysis of the system and equipment grounding arrangements. Section 7 covers the basic functions and requirements for system and equipment grounding. Several grounding arrangements that are applicable to emergency and standby power systems are illustrated in Section 7.

6.10 Conclusions

Protection of emergency and standby power systems determines to a great degree the reliability of power systems. Although the total system should be considered in the protection scheme, overall system protection is only as good as that of its individual components. With emphasis on protecting power supply investment, a designer must be sure that the total investment is not overly compromised. The user must carefully evaluate standard and optional protection schemes supplied with equipment and apply the design best suited to his needs.

6.11 Standards References

The following standards publications were used in the preparation of this section.

ANSI C37.16-1980, Preferred Ratings,

Related Requirements and Application Recommendations for Low-Voltage Power Circuit Breakers and AC Power Circuit Protectors

ANSI/IEEE C37.26-1972, Guide for Methods of Power Factor Measurement for Low-Voltage Inductive Test Circuits (R 1977)

IEEE Std 142-1972, Grounding of Industrial and Commercial Power Systems

IEEE Std 242-1975, Recommended Practice for Protection and Coordination of Industrial and Commercial Power Systems.

IEEE Std 357-1973, Protective Relaying of Utility-Consumer Interconnections

NFPA No 30-1973, Flammable Combustible Liquids Code

NFPA No 70-1978, National Electrical Code

6.12 Bibliography

[1] REICHENSTEIN, H.W., and CASTENSCHIOLD, R. Coordinating Overcurrent Protective Devices with Automatic Transfer Switches in Commercial and Institutional Power Systems. *IEEE Transactions on Industry Applications*, vol TOD-74-17.

[2] IEEE Tutorial Course Text, Surge Protection in Power Systems, 79EHo144-6-PWR.

7. Grounding

7.1 Introduction

7.1.1 *General.* An ultimate goal of system protection is to prevent ground faults from occurring by proper design, installation, operation, and maintenance of electrical equipment and systems. However, some probability of ground faults due to accidents or insulation failure will exist in any electrical installation. Therefore, protection must be provided to safely clear the line-to-ground faults that may occur.

Reliable ground-fault protection for low voltage electrical systems requires coordinated design, proper installation, and routine maintenance of the following:

(1) Circuit protective equipment
(2) System grounding
(3) Equipment grounding

All references to the NEC in this section refer to the 1978 edition of the National Electrical Code.

7.1.2 *Circuit Protective Equipment.* Different types of circuit protection are satisfactory for ground-fault protection. Some commonly used types are fuses, circuit breakers with series trips, and circuit interrupting equipment tripped by ground-fault current sensing devices. Selection and application of circuit protective equipment requires a detailed analysis of each system and circuit to be protected, including the system and equipment grounding arrangements.

7.1.3 *System and Equipment Grounding.* This section presents basic system and equipment grounding requirements and grounding arrangements for emergency and standby power systems rated 600 V or less.

Factors which influence the selection of the type of system grounding, and the fundamental design principles for system and equipment grounding, are covered in detail in IEEE Std 141-1976 (IEEE Red Book) and IEEE Std 142-1972 (IEEE Green Book). This section is not intended to recommend the type of system grounding or the type of transfer equipment that should be selected for emergency and standby power systems. The purpose of the section is to define the fundamental grounding requirements and to illustrate grounding arrangements for several types of system grounding and transfer schemes that are most often selected for emergency and standby power systems in industrial and commercial installations.

7.2 System and Equipment Grounding Functions

7.2.1 *General*. System grounding pertains to the manner in which a circuit conductor of a system is intentionally connected to earth, or to some conducting body which is effectively connected to earth or serves in place of earth. The following types of system grounding are discussed in this section:

(1) *Solidly·Grounded*. A grounding electrode conductor connects the grounded conductor of the system to the grounding electrode(s) and no impedance is intentionally inserted in the connection.

(2) *Resistance Grounded*. A grounding resistor is inserted in the connection between the grounded conductor of the system and the grounding electrode(s).

(3) *Ungrounded*. None of the circuit conductors of the system are intentionally grounded.

Equipment grounding is the bonding together of all conductive enclosures for conductors and equipment in each circuit with equipment grounding conductors. These equipment grounding conductors are required to run with or enclose the circuit conductors, and they provide a permanent, low-impedance conductive path for ground-fault current. In solidly grounded systems, the equipment grounding conductors are bonded to the grounded circuit conductor and to the system grounding conductor(s) at specific points as shown in Figs 57 and 58.

7.2.2 *System Grounding Functions*.

**Fig 57
System and Equipment Grounding for
Solidly Grounded, Service-Supplied System**

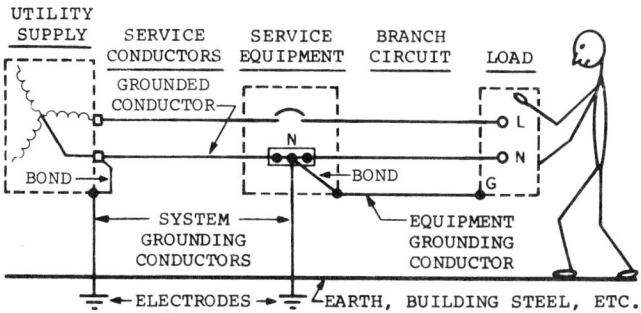

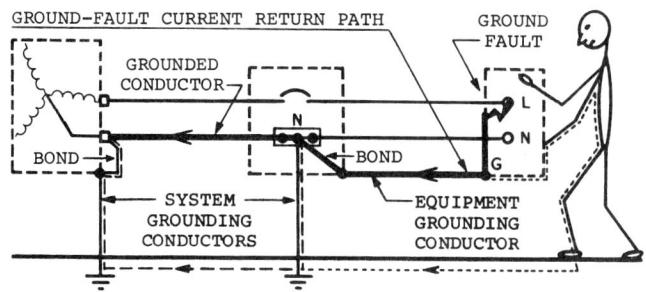

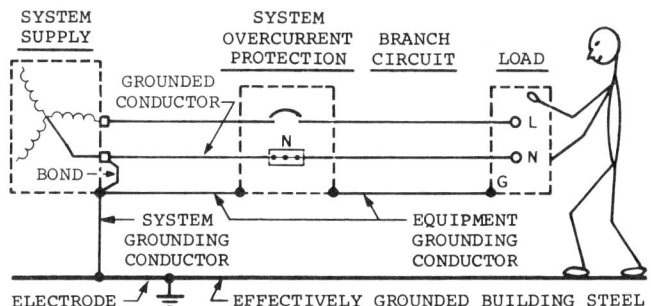

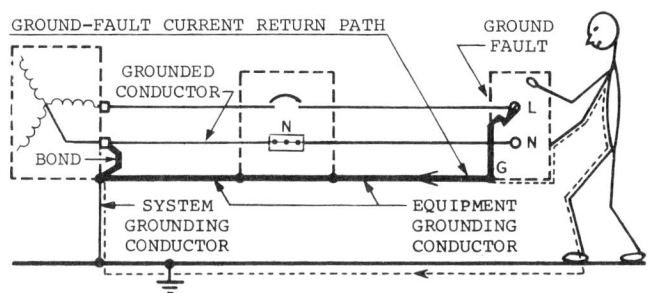

Fig 58
System and Equipment Grounding for
Solidly Grounded, Separately Derived System

Systems and circuit conductors are grounded to limit voltages due to lightning, line surges, or unintentional contact with higher voltage lines and to stabilize the voltage to ground during normal operation. Systems and circuit conductors are solidly grounded to facilitate overcurrent device operation in case of ground faults.

Grounding electrodes and the grounding electrode conductors that connect the electrodes to the system grounded conductor are not intended to conduct ground-fault currents that are due to ground faults in equipment, raceways, or other conductor enclosures. In solidly grounded systems, the ground-fault current flows through the equipment grounding conductors from a ground fault anywhere in the system to the bonding jumper between the equipment grounding conductors and the system grounded conductor as illustrated in Figs 58 and 59.

In solidly grounded service-supplied systems, the ground-fault current return path is completed through the bonding jumper in the service equipment and the grounded service conductor to the supply transformer as shown in Fig 58.

7.2.3 *Equipment Grounding Functions.* Equipment grounding systems, consisting of interconnected networks of equipment grounding conductors, perform the following basic functions:

(1) They limit the voltage to ground (shock voltage) on the exposed non-current-carrying metal parts of equipment, raceways, and other conductor enclosures in case of ground faults.

(2) They safely conduct ground-fault

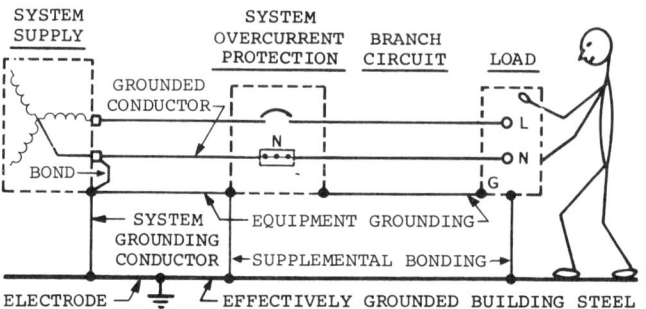

Fig 59
Supplemental Equipment Bonding
for Separately Derived System

currents of sufficient magnitude for fast operation of the circuit protective devices.

In order to ensure the performance of the above basic functions, equipment grounding conductors should:

(1) Be permanent and continuous

(2) Have ample capacity to conduct safely any ground-fault current likely to be imposed on them

(3) Have impedance sufficiently low to limit the voltage to ground to a safe magnitude and to facilitate the operation of the circuit protective devices.

As illustrated in Figs 57 and 58, a person contacting a conductive enclosure in which there is a ground fault will be protected from shock injury if the equipment grounding conductors provide a shunt path of sufficiently low impedance to limit the current through the person's body to a safe magnitude. In solidly grounded systems, the ground-fault current actuates the circuit protective devices to automatically deenergize a faulted circuit and remove the shock exposure.

The same network of equipment grounding conductors must be provided for solidly grounded systems, high resistance grounded systems, and ungrounded systems. Equipment grounding conductors are required in high resistance grounded and ungrounded systems to provide shock protection and to present a low-impedance path for phase-to-phase fault currents in case the first ground fault is not located and cleared before another ground fault occurs on different phase in the system.

7.3 Supplemental Equipment Bonding

Exposure to electrical shock can be reduced by additional supplemental equipment bonding between the conductive enclosures for conductors and equipment and adjacent conductive materials.

The supplemental equipment bonding shown in Fig 59 contributes to equalizing the potential between exposed non-current-carrying metal parts of the electrical system and adjacent grounded building steel when ground faults occur. The inductive reactance of the ground-fault circuit will normally prevent a significant amount of ground-fault current from flowing through the supplemental bonding connections.

Ground-fault current will flow through the path that provides the lowest ground-fault circuit impedance. The ground-fault current path that minimizes the inductive reactance of the ground-fault circuit is through the equipment grounding conductors which are required to run with or enclose the circuit conductors. Therefore, practically all of the ground-fault current will flow through the equipment grounding conductors, and the ground-fault current through the supplemental bonding connections will be no more than required to equalize the potential at the bonding locations.

7.4 Objectionable Current Through Grounding Conductors

System and equipment grounding conductors should be installed and connected in a manner that will prevent an objectionable flow of current through the grounding conductors or grounding paths. If a grounded (neutral) circuit conductor is connected to the equipment grounding conductors at more than one point, or if it is grounded at more than one point, stray neutral current paths will be established. Stray neutral currents that flow through paths other than the intended grounded (neutral) circuit conductors during normal operation of a system will be objectionable if they contribute to any of the following:

(1) Interference with the proper operation of equipment, devices, or systems that are sensitive to electromagnetic interference such as electronic equipment, communications systems, computer systems, etc

(2) Interference with the proper sensing and operation of ground-fault protection equipment

(3) Arcing of sufficient energy to ignite flammable materials

(4) Detonation of explosives during production, storage, or testing

(5) Overheating due to heat generated in raceways, etc, as a result of stray current.

The grounded service conductor of service-supplied systems should be grounded at the service equipment. If the supply transformer is located outside of the building, the grounded service conductor should also be grounded on the secondary side of the supply transformer, either at the transformer or elsewhere ahead of the service equipment. A grounding connection should be made to any grounded (neutral) circuit conductor on the load side of the service disconnecting means.

The stray neutral current illustrated in Fig 60 is due to multiple grounding of the grounded service conductor. Where there

**Fig 60
Stray Neutral Current Due to Multiple Grounding of Grounded Service Conductor**

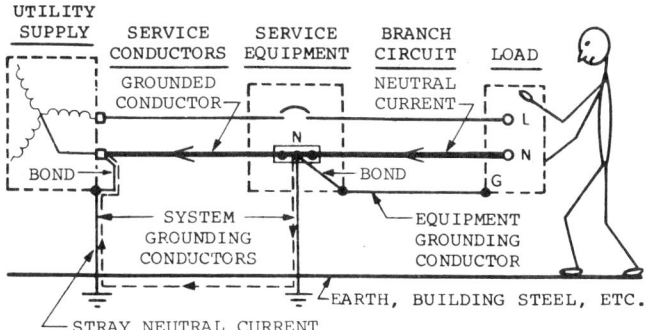

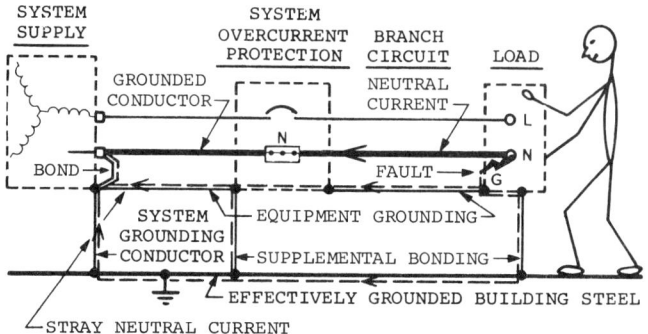

**Fig 61
Stray Neutral Current Due to Unintentional
Grounding of a Grounded Circuit Conductor**

is a single overhead service drop connected to only one set of service entrance conductors, the stray neutral current illustrated in Fig 60 will normally be of insufficient magnitude to be objectionable.

Objectionable stray neutral currents are frequently caused by unintentional neutral-to-ground faults as shown in Fig 61. Neutral-to-ground faults are difficult to locate, but should be suspected if there are objectionable stray neutral currents.

7.5 System Grounding Requirements

AC systems of 50 to 1000 V should be solidly grounded where the maximum voltage to ground on the ungrounded conductors will not exceed 150 V. Systems that supply phase-to-neutral loads should also be solidly grounded. The following commonly used systems are required to be solidly grounded:

(1) 120/240 V single-phase, three-wire
(2) 208Y/120 V three-phase, four-wire
(3) 480Y/277 V three-phase, four-wire
(4) 240/120 V three-phase, four-wire delta (midpoint of one phase used as a grounded circuit conductor)

The following commonly used systems are not required to be solidly grounded:

(1) 240 V delta, three-phase, three-wire
(2) 480 V three-phase, three-wire
(3) 600 V three-phase, three-wire

Where phase-to-neutral loads need not be served, there is a growing trend toward high resistance grounding for 480 V and 600 V three-phase systems in industrial establishments. High resistance system grounding combines some of the advantages of solidly grounded systems and ungrounded systems. System overvoltages are held to acceptable levels during ground faults, and the potentially destructive effects of high magnitude ground-fault currents that occur in high capacity, solidly grounded systems are eliminated. Ground-fault current is limited by the system grounding resistor to a magnitude that permits continued operation of a system while a ground fault is located and cleared. However, if a ground fault is not located and cleared before another ground fault occurs on another phase in the system, the high magnitude phase-to-phase fault current will flow through the equipment grounding conductors and operate the circuit protective equipment. Factors that influence selecting the type of system grounding are cov-

GROUNDING

ered in the standards referenced in 7.1.3.

7.6 Types of Equipment Grounding Conductors

Equipment grounding conductors represented in the diagrams in this section may be any of the types permitted in NEC Section 250-91(b) if this section is adopted by the authority having jurisdiction. Where raceway, cable tray, cable armor, etc, are used as equipment grounding conductors, all joints and fittings must be made tight to provide an adequate conducting path for ground-fault currents.

Earth and the structural metal frame of a building may be used for supplemental equipment bonding, but they should not be used as the sole equipment grounding conductor for ac systems. Equipment grounding conductors for ac systems should run with or enclose the conductors of each circuit.

Where copper or aluminum wire is used as equipment grounding conductors for circuits having paralleled conductors in multiple metal raceways, an equipment grounding conductor should be run in each raceway. The size of each paralleled equipment grounding conductor is a function of the rating of the circuit overcurrent protection.

7.7 Grounding for Separately Derived and Service-Supplied Systems

Basic grounding connections for solidly grounded separately derived and service-supplied systems are illustrated in Fig 62 with reference to NEC sections in which the grounding requirements are specified.

The grounding requirements for separately derived systems and service-supplied systems are similar, but there are three important differences:

(1) The system grounded conductor for a separately derived system should be grounded at only one point. The single system grounding point is specified as the source of the separately derived system and ahead of any system disconnecting means or overcurrent devices. Where the main system disconnecting means is adjacent to the generator or transformer supplying a separately derived system, the grounding connection to the system grounded conductor may be made at, or ahead of, the system disconnecting means.

The system grounded conductor for a service-supplied system should be grounded at the service equipment and elsewhere on the secondary of the transformer that supplies the service, if the supply transformer is not in the same building as the service equipment.

(2) The preferred grounding electrode for a separately derived system is the nearest effectively grounded structural metal member of the structure or the nearest effectively grounded water pipe. The grounding electrode system for a service-supplied system should be in accordance with established code requirements.

(3) In solidly grounded separately derived systems, the equipment grounding conductors should be bonded to the system grounded conductor and to the grounding electrode conductor at or ahead of the main system disconnecting means or overcurrent device. The equipment grounding conductor should always be connected to the enclosure of the supply transformer or generator, as illustrated in Fig 62.

In solidly grounded service-supplied systems, the equipment grounding conductors should be bonded to the system grounded conductor and to the grounding electrode conductor at the service equipment. The grounded service conductor may be used to ground the noncurrent-

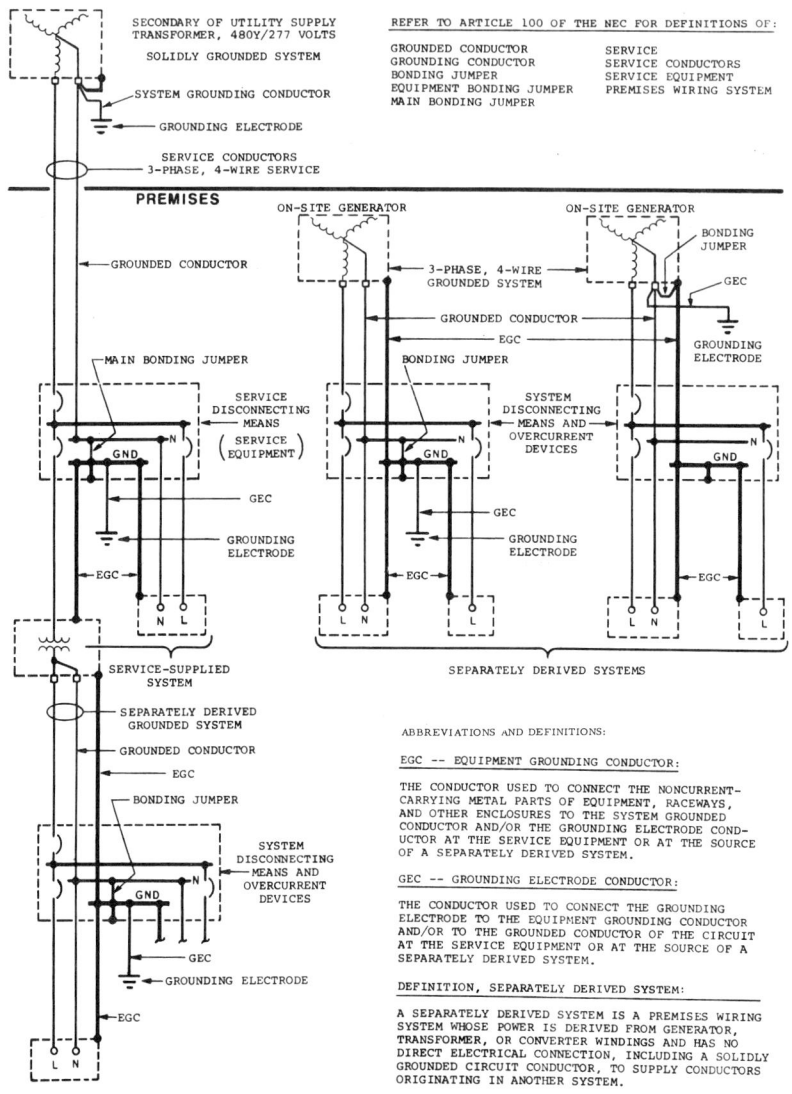

**Fig 62
System and Equipment Grounding for
Separately Derived and Service-Supplied Systems**

carrying metal parts of equipment on the supply side of the service disconnecting means, and the grounded service conductor may also serve as the ground-fault current return path from the service equipment to the transformer that supplies the service.

7.8 Grounding Arrangements for Emergency and Standby Power Systems

A primary consideration in designing emergency and standby power systems is to satisfy the user's needs for continuity of electrical service. The type of system grounding that is employed, and the arrangement of system and equipment grounding conductors, will affect the service continuity. Grounding conductors and connections must be arranged so that objectionable stray neutral currents will not exist and ground-fault currents will flow in low-impedance, predictable paths which will protect personnel from electrical shock and assure proper operation of the circuit protective equipment.

Where phase-to-neutral loads must be served, systems are required to be solidly grounded. However, 600 V and 480 V systems may be high resistance grounded or ungrounded where a grounded circuit conductor is not used to supply phase-to-neutral loads. High resistance grounded or ungrounded systems may provide a higher degree of service continuity than solidly grounded systems.

This section discusses grounding arrangements for emergency and standby power systems that are solidly grounded, high resistance grounded, and ungrounded.

7.9 Systems with a Grounded Circuit Conductor

Where grounded (neutral) conductors are used as circuit conductors in systems that have emergency or standby power supplies, the grounding arrangement must be carefully planned to avoid objectionable stray currents. For example, stray neutral currents and ground-fault currents in unplanned, undefined conducting paths may cause serious sensing errors by ground-fault protection equipment. A precautionary note is included in NEC Section 230-95 which states, "Where ground-fault protection is provided for the service disconnecting means and interconnection is made with another supply system by a transfer device, means or devices may be needed to assure proper ground-fault sensing by the ground-fault protection equipment."

7.9.1 *Solidly Interconnected Multiple-Grounded Neutral.* A grounded (neutral) circuit conductor is permitted to be solidly connected (not switched) in the transfer equipment. Therefore, a neutral conductor is permitted to be solidly interconnected between a service-supplied normal source and an on-site generator which serves as an emergency or standby source as illustrated in Fig 63.

Grounding connections to the grounded (neutral) conductor on the load side of the service disconnecting means is not recommended, so the grounding connection to the generator neutral in Fig 63 *should not be made*. Such multiple grounding of the neutral circuit conductor may cause stray currents that are likely to be objectionable and will cause ground-fault current to flow in paths that may adversely affect the operation of ground-fault protection equipment.

(1) *Fig 63:* The grounding connection to the neutral conductor at the on-site generator in Fig 63 is on the load side of the service disconnecting means, and is not recommended and may not satisfy code requirements. Where the grounded (neutral) conductor is solidly connected (not switched) in the transfer equipment, the system supplied by the on-site generator should not be considered as a separately derived system.

(2) *Fig 64:* The grounding connection to the neutral conductor at the on-site generator in Fig 64 completes a conducting path for stray neutral current. The magnitude of the stray current will be a function of the relative impedances of the neutral current paths. Where ground-fault protection is provided at the service disconnecting means, the stray neutral current may adversely affect the operation of the ground-fault protection equipment.

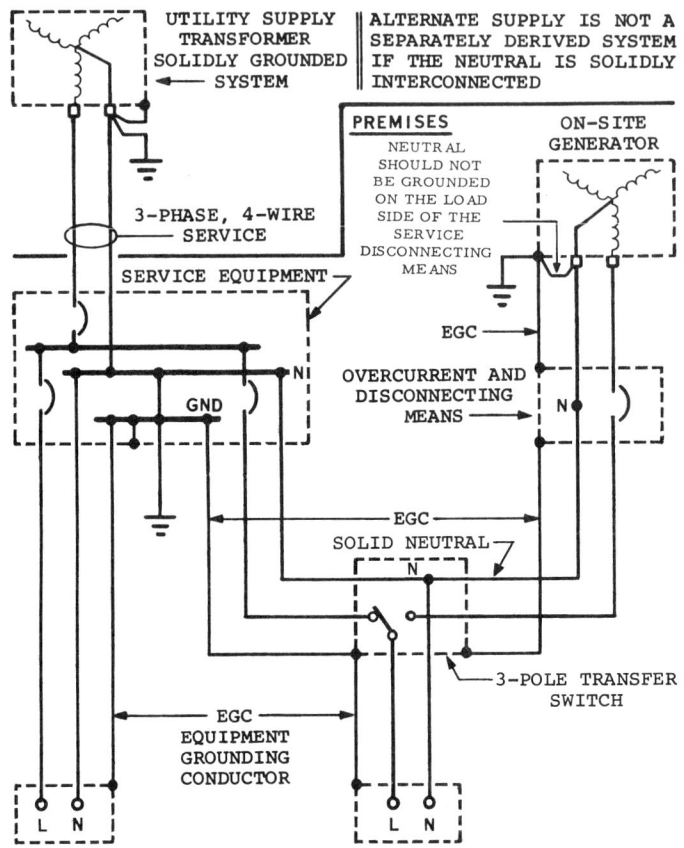

Fig 63
Solidly Interconnected Neutral Conductor
Grounded at Service Equipment and
at Source of Alternate Power Supply

(3) *Fig 65:* The multiple grounding connections to the solidly interconnected neutral in Fig 65 permit a portion of the ground fault current returning to the normal source to flow through the neutral grounding bond at the on-site generator and bypass the sensor for the ground-fault protection at the service equipment. The main service disconnecting means is not supposed to trip for a ground fault on a feeder or branch circuit, so this is not a serious problem provided the feeder to the transfer switch does not have ground-fault protection equipment that is actuated by a ground-fault current sensor. However, ground-fault protection for feeders may be required in health-care facilities if the service disconnecting means is equipped with ground-fault protection.

Where ground-fault protection is applied to a feeder from the service equipment to the transfer equipment, the ground-fault current illustrated in Fig 65 could not be accurately detected by a zero sequence ground-fault current sensor for the feeder.

(4) *Fig 66:* The multiple grounding

GROUNDING

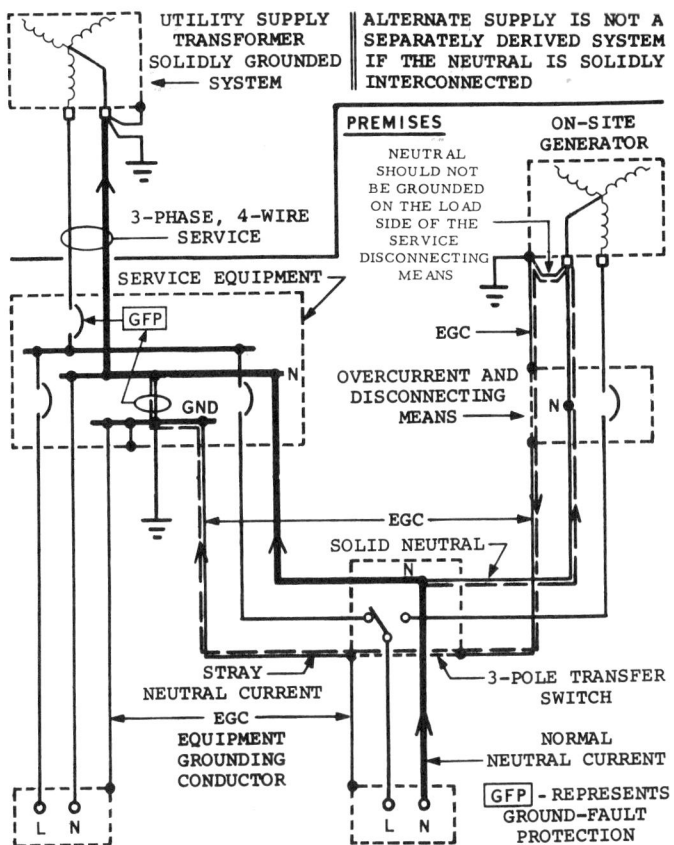

Fig 64
Stray Neutral Current Due to Grounding
the Neutral Conductor at Two Locations

connections to the solidly interconnected neutral in Fig 66 permit a portion of the ground-fault current returning to the on-site generator to pass through the sensor for the ground-fault protection at the service equipment. This arrangement could result in tripping the service disconnecting means by ground-fault current supplies from the on-site generator.

7.9.2 *Neutral Conductor Transferred by Transfer Means.* The transfer switch may have an additional pole for switching the neutral conductor, or the neutral may be transferred by make-before-break overlapping neutral contacts in the transfer switch. Where the neutral circuit conductor is transferred by the transfer equipment, an emergency or standby system supplied by an on-site generator is a separately derived system. A separately derived system with a neutral circuit conductor should be solidly grounded at or ahead of the system disconnecting means.

(1) *Fig 67:* The neutral conductor between the service equipment and the on-site generator in Fig 67 is completely isolated by the transfer switch and is solidly grounded at the service equipment and at

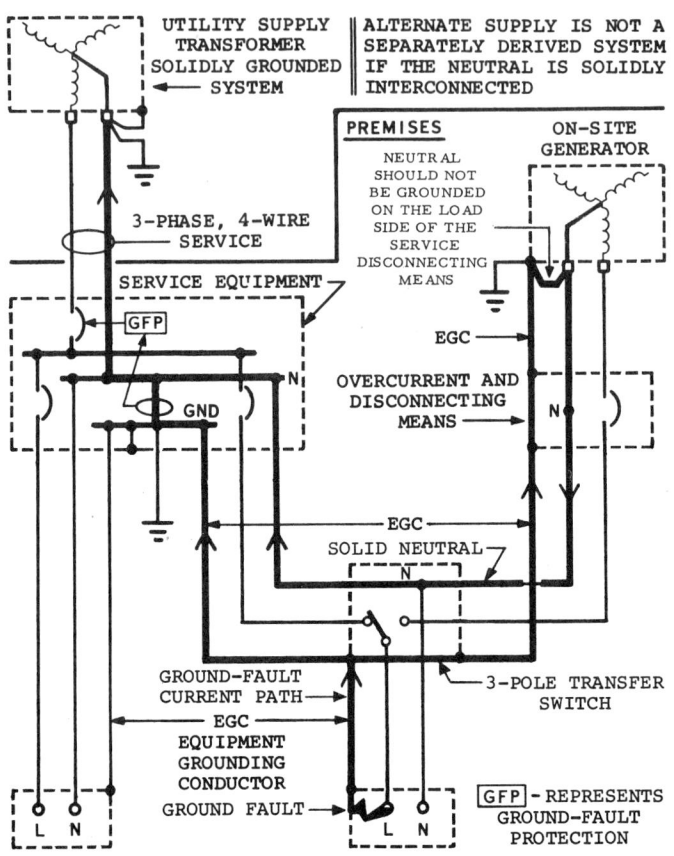

Fig 65
Ground-Fault Current Return Paths to Normal Supply, Neutral Grounded at Two Locations

the on-site generator. Where the neutral circuit conductor is switched by the transfer equipment, the stray neutral current paths and the undesirable ground-fault current paths illustrated in Figs 64, 65, and 66 are eliminated.

The normal supply and the alternate supply in Fig 67 are equivalent to two separate radial systems because all of the circuit conductors from both supplies are switched by the transfer switch. Since both systems are completely isolated from each other and are solidly grounded, ground-fault sensing and protection can be applied to the circuits of the normal source and the emergency or standby source as it is applied in a single radial system.

(2) *Fig 68:* This diagram shows that unintentional neutral grounds will cause stray neutral currents that may be objectionable. Therefore, in addition to carefully planning the intentional grounding connections, systems should be kept free of unintentional grounds on the neutral circuit conductors.

(3) *Fig 69:* This diagram is similar to Fig 67 except there are two on-site

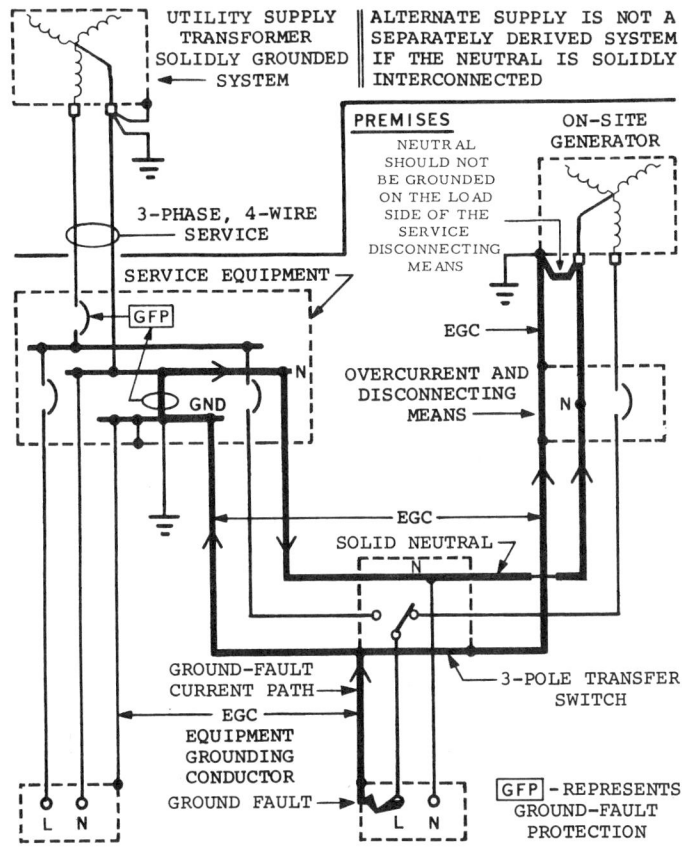

Fig 66
Ground-Fault Current Return Paths to Alternate
Supply, Neutral Grounded at Two Locations

generators connected for parallel operation and the neutral conductor is grounded at the generator switchgear instead of at the generator. Where the generator switchgear is adjacent to the generators, the grounding connections may be made in the generator switchgear, as illustrated in Fig 69.

Switching of the neutral conductor permits grounding of the neutral at the generator location. This may be desirable for the following reasons:

(1) An engine-generator set is often remotely located from the grounded utility service entrance, and the ground potentials of the two locations may not be the same.

(2) Good engineering practice requires the automatic transfer switch to be located as close to the load as possible to provide maximum protection against cable or equipment failures within the facility. The distance of cable between incoming service and the transfer switch and then to the engine-generator set may be substantial. Should complete cable failure occur with the neutral conductor not grounded at the generator location,

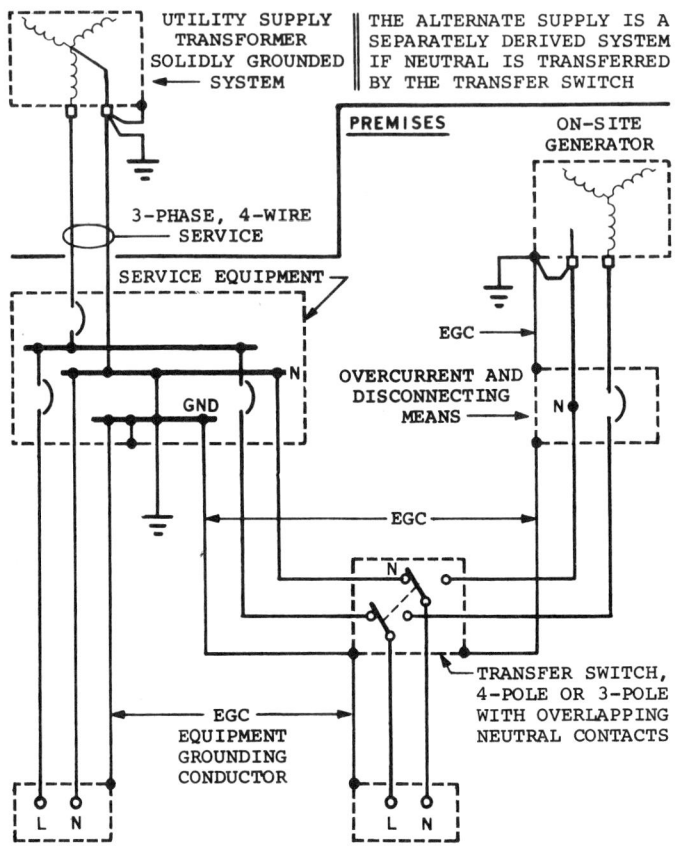

Fig 67
**Transferred Neutral Conductor
Grounded at Service Equipment and
at Source of Alternate Power Supply**

the load would be transferred to an ungrounded emergency power system. This could jeopardize emergency service continuity and possibly lead to additional failures. Concurrent failure of equipment (breakdown between line and equipment ground) after transfer to emergency may not be detected. Thus the generator frame may approach line potential, causing a substantial voltage difference between the generator frame and the neutral conductor.

(3) Some local codes require ground-fault protection while the engine-generator is operating. This may present a sensing problem if the neutral conductor of the generator is not connected to a grounding electrode at the generator site and proper isolation of neutrals is not provided.

(4) When the transfer switch is in the emergency position, other problems may occur if the engine-generator set is not properly grounded. A ground-fault condition could cause nuisance tripping of the normal source circuit breaker even though load current is not flowing through the breaker. Furthermore, both

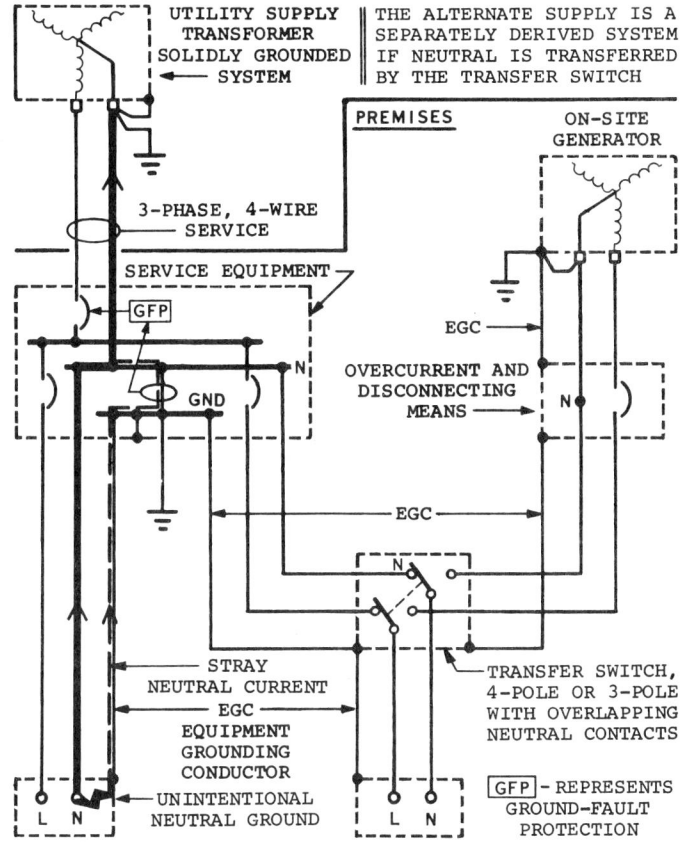

Fig 68
Stray Neutral Current Due to
Unintentionally Grounded Neutral Conductor

the normal neutral conductor and the emergency neutral conductor would be simultaneously vulnerable to the same ground-fault current. Thus a single fault could jeopardize power to critical loads even though both utility and emergency power are available. Such a condition may be in violation of codes requiring independent wiring and separate emergency feeders.

7.9.3 *Neutral Conductor Isolated by a Transformer.* Where a transferable load is supplied by a system that is derived from an on-site isolating transformer and the transfer equipment is ahead of the transformer, as illustrated in Fig 70, a grounded (neutral) circuit conductor is not required from either the normal or alternate supply to the transformer primary. The isolating transformer permits phase-to-neutral transferable loads to be supplied without a grounded (neutral) circuit conductor in the feeders to the transfer switch.

(1) *Fig 70:* The system supplied by the isolating transformer in Fig 70 is a separately derived system, and if it is required to be solidly grounded, it should be

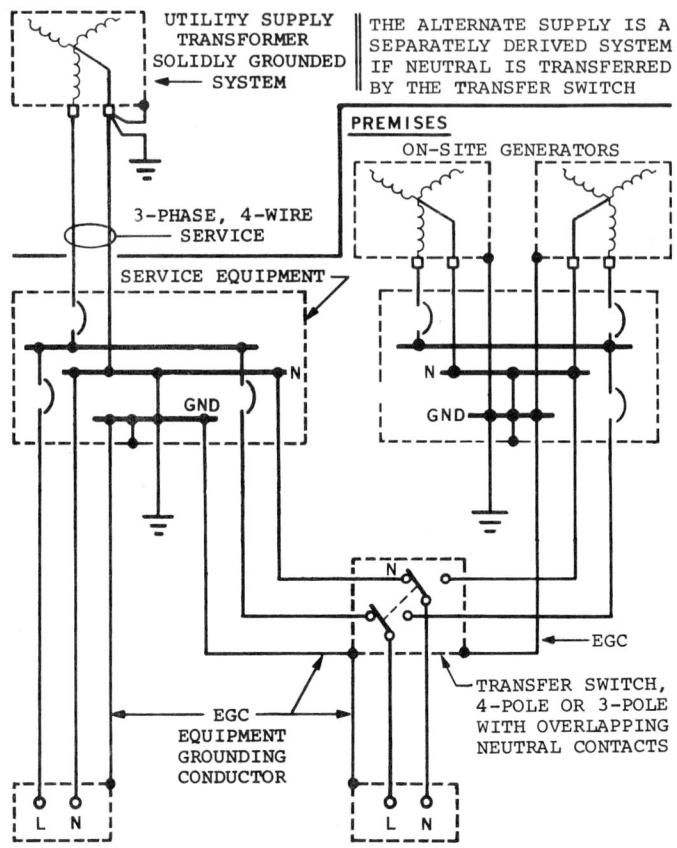

Fig 69
Transferred Neutral Conductor Grounded at Service Equipment and at Switchgear for Two On-Site Generators Connected in Parallel

grounded in accordance with code requirements. The neutral circuit conductor for the transferable load in Fig 70 is supplied from the secondary of the isolating transformer.

If the on-site generator in Fig 70 is rated 480Y/277 V or 600Y/347 V, its neutral may not need to be solidly grounded because the neutral is not used as a circuit conductor. Therefore, the type of system grounding for such a generator is optional. However, the generator frame should be solidly grounded whether its neutral is ungrounded, high resistance grounded, or solidly grounded.

(2) *Fig 71:* This diagram illustrates ground-fault current return paths where the grounded (neutral) circuit conductor of a transferable load is isolated by a transformer. Any stray neutral current or ground-fault current on the secondary of the isolating transformer will have no effect on ground-fault protection equipment at the service equipment or at the generator.

7.9.4 *Solidly Interconnected Neutral*

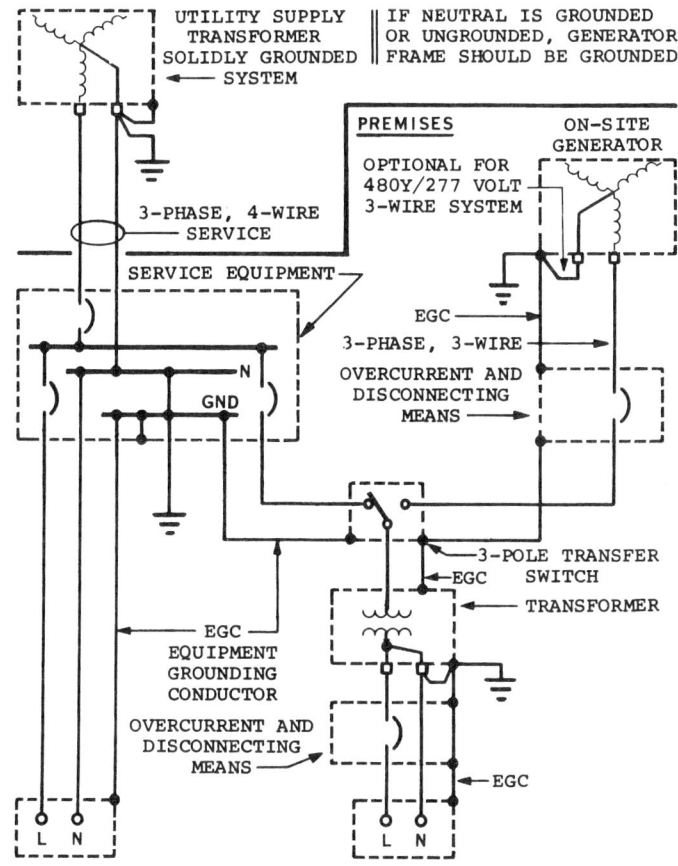

Fig 70
Solidly Grounded Neutral Conductor for
Transferred Load Isolated By Transformer

Conductor Grounded at Service Equipment Only. Where the grounded (neutral) circuit conductor is solidly connected (not switched) in the transfer equipment, an emergency or standby system supplied by an on-site generator should not be considered a separately derived system. The solidly interconnected grounded (neutral) conductor need only to be grounded at the service equipment.

(1) *Fig 72:* The solidly interconnected neutral conductor in Fig 72 is grounded at the service equipment only and there are no conducting paths for stray neutral currents. If the generator is not adjacent to the service equipment, the generator frame should be connected to a grounding electrode such as effectively grounded building steel.

(2) *Fig 73:* Where the solidly interconnected neutral conductor is grounded at the service equipment only as shown in Fig 73, the ground-fault current return path from the transferable load to the on-site generator is through the equipment grounding conductor from the transfer

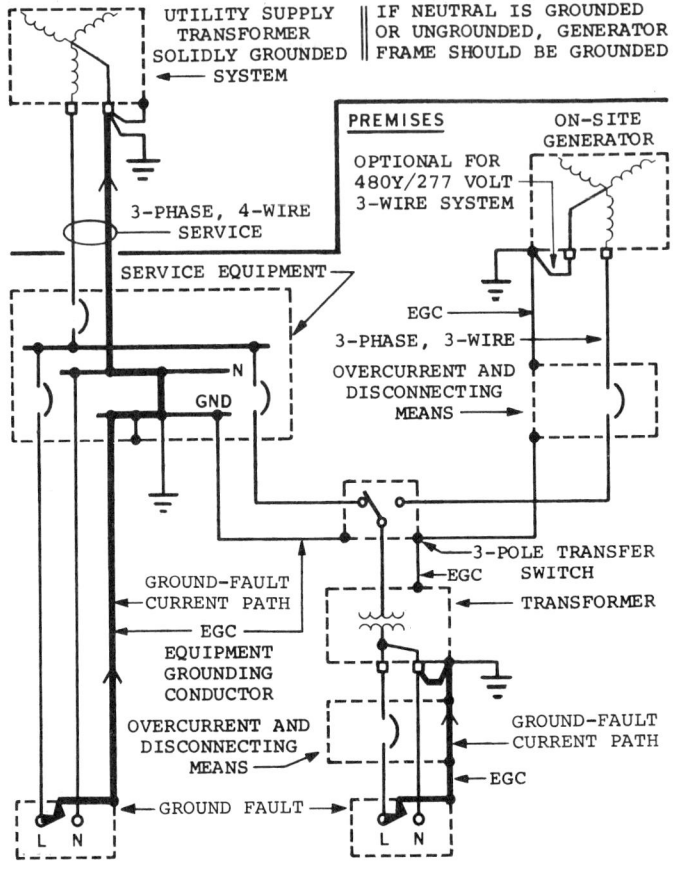

Fig 71
Ground-Fault Current Return Paths
Transferred Load Isolated By Transformer

switch to the service equipment, thence through the main bonding jumper in the service equipment and the neutral conductor from the service equipment to the generator.

The ground-fault current illustrated in Fig 73 might trip the service disconnecting means even though the ground fault is on a circuit supplied by the generator. A signal could be derived from a ground-fault sensor on the generator neutral conductor to block the ground-fault protection equipment at the service in case of ground faults while the system is transferred to the generator. Such blocking signals require careful analysis to ensure proper functioning of the ground-fault protection equipment.

If the neutral conductor between the service equipment and the transfer switch in Fig 73 is intentionally or accidentally disconnected, the generator will be ungrounded. Therefore, the integrity of the neutral conductor must be maintained from the service equipment to the transfer switch while the load is transferred t

GROUNDING

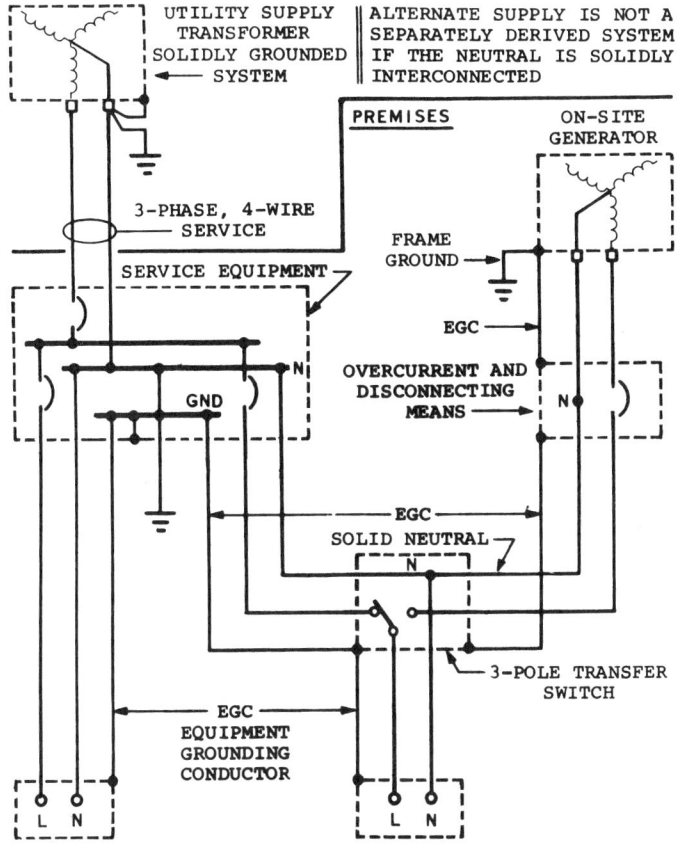

**Fig 72
Solidly Interconnected Neutral Conductor
Grounded at Service Equipment Only**

the generator. The equipment grounding conductor must also be maintained from the service equipment to the transfer equipment in order to provide a ground-fault current return path from the transferable load to the generator.

For increased reliability, multiple transfer switches, located close to the loads, are often used rather than one transfer switch for the entire load. Therefore, consideration should be given to the possibility of cable or equipment failure between the service equipment and the transfer switches, thus possibly causing an ungrounded emergency or standby power system.

7.10 Systems Without a Grounded Circuit Conductor

Three-phase, three-wire, 480 V and 600 V systems, which are extensively used in industrial establishments, do not require the use of grounded conductors as circuit conductors. There are more system grounding options where emergency and standby power systems do not require a

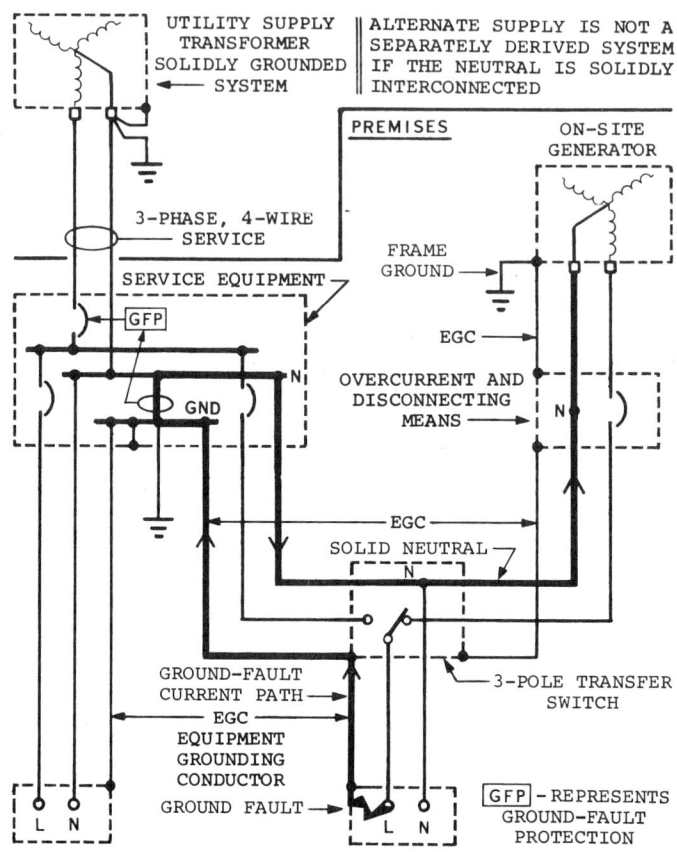

**Fig 73
Ground-Fault Current Return Path to
Alternate Supply, Neutral Conductor
Grounded at Service Equipment Only**

grounded circuit conductor to supply phase-to-neutral loads.

7.10.1 *Solidly Grounded Service.* In many installations the service to the premises will be solidly grounded, three-phase, four-wire where a grounded (neutral) circuit conductor is not required for loads that are provided with an on-site emergency or standby supply. An on-site emergency or standby supply is not always required to have the same type of system grounding as the normal supply to the premises.

(1) *Fig 74:* The generator in Fig 74 supplies a three-phase, three-wire system. If the generator is rated 480Y/277 V or 600Y/347 V, it need not be solidly grounded because its neutral is not used as a circuit conductor.

An on-site generator that is not required to be solidly grounded may be high resistance grounded or ungrounded. A high resistance grounded or ungrounded emergency or standby power supply provides a high degree of service continuity because the circuit protective equipment

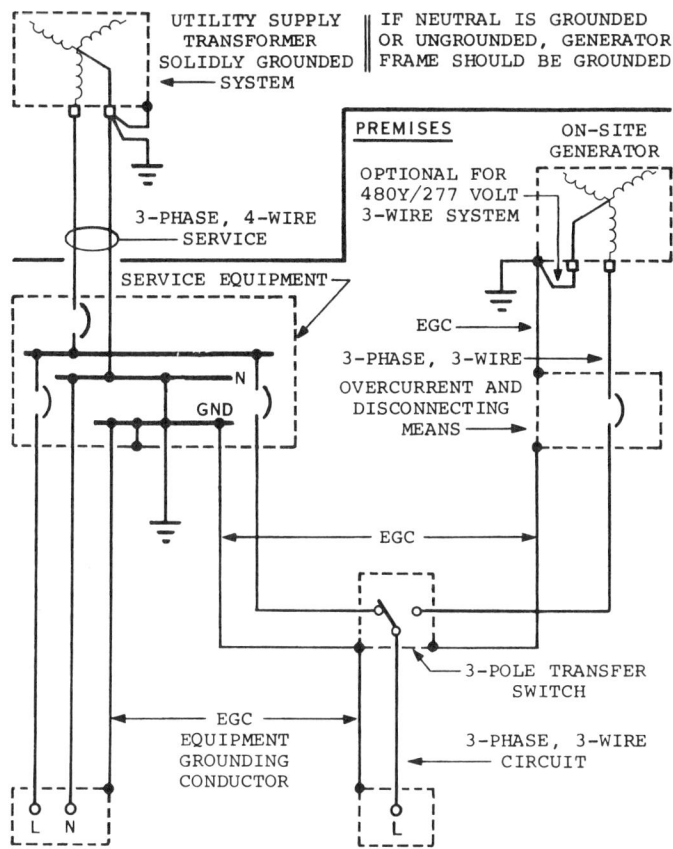

Fig 74
Three Pole Transfer Switch for Transfer
to an Alternate Power Supply Without
a Grounded Circuit Conductor

will not be tripped by the first ground fault on the system.

If the generator in Fig 74 is solidly grounded, it must be grounded at or ahead of the generator disconnecting means.

(2) *Fig 75:* Interlocked circuit breakers are used as the transfer means in Fig 75. The grounding arrangement in Figs 74 and 75 are the same. Where a grounded (neutral) circuit conductor is not required, the type of transfer equipment that is employed is not a consideration in selecting the type of system grounding for the emergency or standby supply.

Interlocked circuit breakers should not be selected as a transfer means without considering additional means to isolate the normal and alternate circuit conductors and equipment for maintenance work. The overall reliability of a system may be reduced due to additional exposure where the circuit conductors from the normal supply are solidly connected to the circuit conductors from the emergency or standby supply. The circuit wiring for

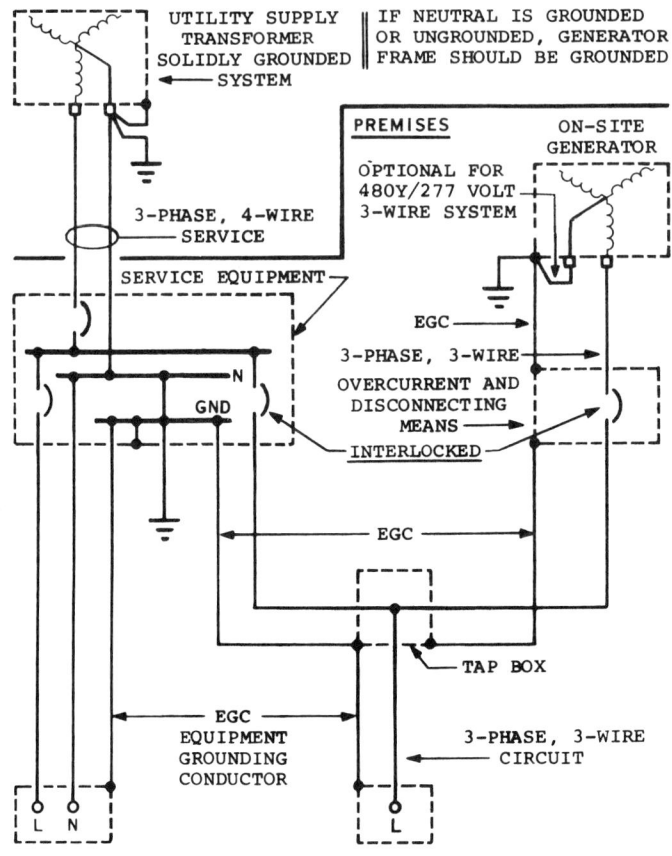

Fig 75
Interlocked Circuit Breakers for Transfer
to an Alternate Power Supply Without
a Grounded Circuit Conductor

an emergency system should be kept entirely independent of all other wiring or equipment except in transfer switches or in junction boxes and in fixtures for exit and emergency lighting.

(3) *Fig 76:* A 480 V or 600 V, three-phase, three-wire, on-site generator that is high resistance grounded may serve as an emergency or standby power supply for a three-phase, three-wire system that is normally supplied by a solidly grounded service, as illustrated in Fig 76.

7.10.2 *High Resistance Grounded Service.* Where the three-phase, three-wire critical load is relatively large compared with loads that require a grounded (neutral) circuit conductor, a high resistance grounded service with a high resistance grounded emergency or standby power supply is sometimes considered. This arrangement requires an on-site transformer for loads that require a neutral circuit conductor. If an on-site transformer is provided for emergency lighting, it must be supplied from the normal service and the emergency supply

GROUNDING

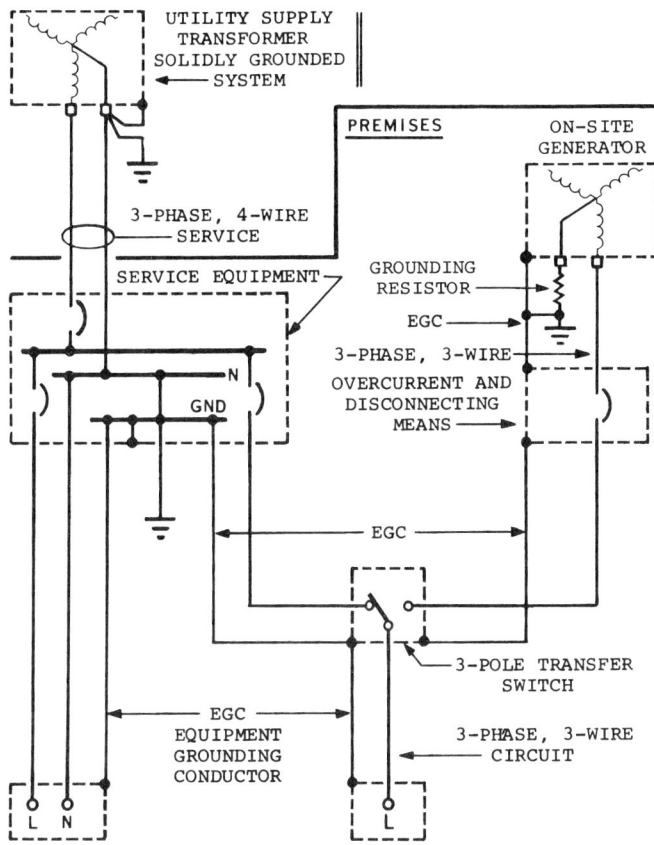

Fig 76
High Resistance Grounded Alternate Power
Supply Without a Grounded Circuit Conductor.

through automatic transfer equipment.

(1) *Fig 77:* The supply transformer for the normal service and the on-site generator in Fig 77 are both high resistance grounded. There are no provisions in this diagram to supply phase-to-neutral loads.

(2) *Fig 78:* The ground-fault current return path to the on-site generator in Fig 78 is completed through the grounding resistor. The grounding resistor limits the line-to-ground fault current to a magnitude that can be tolerated for a time, allowing the ground fault to be located and removed from the system.

High resistance grounded systems should not be used unless they are equipped with ground-fault indicators or alarms, or both, and qualified persons are available to quickly locate and remove ground faults. If ground faults are not promptly removed, the service reliability will be reduced.

7.11 Mobile Engine Generator Sets

The basic requirements for system

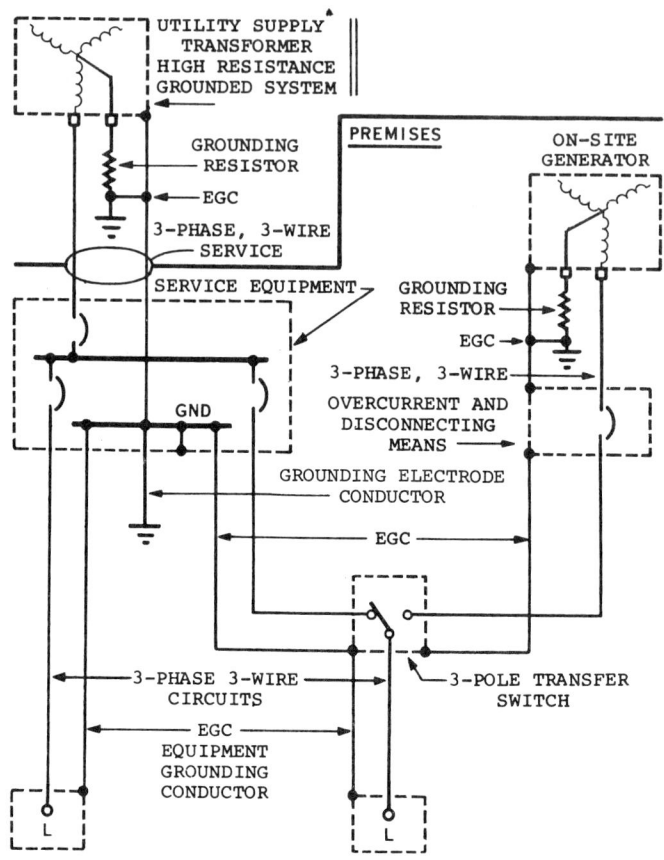

**Fig 77
High Resistance Grounded Systems
for Normal Service and Alternate Supply**

grounding and equipment grounding that are discussed and illustrated in previous sections also apply to mobile engine generator sets when they are used to supply emergency or standby power for systems rated 600 V or less. Grounding arrangements that are acceptable for emergency or standby power supplied by fixed generators are also acceptable where such systems are supplied from mobile engine generators. If ground-fault protection equipment is used for the normal service or for a mobile generator that supplies emergency or standby power, the system and equipment grounding connections must be arranged to prevent stray neutral currents and ground-fault currents in paths that cause improper operation of the ground-fault protection devices.

Where mobile generators are used to supply emergency or standby power for systems similar to Fig 67 in which the neutral is grounded at the generator, it may be desirable to install permanent, pretested system grounding electrodes at

GROUNDING

IEEE
Std 446-1980

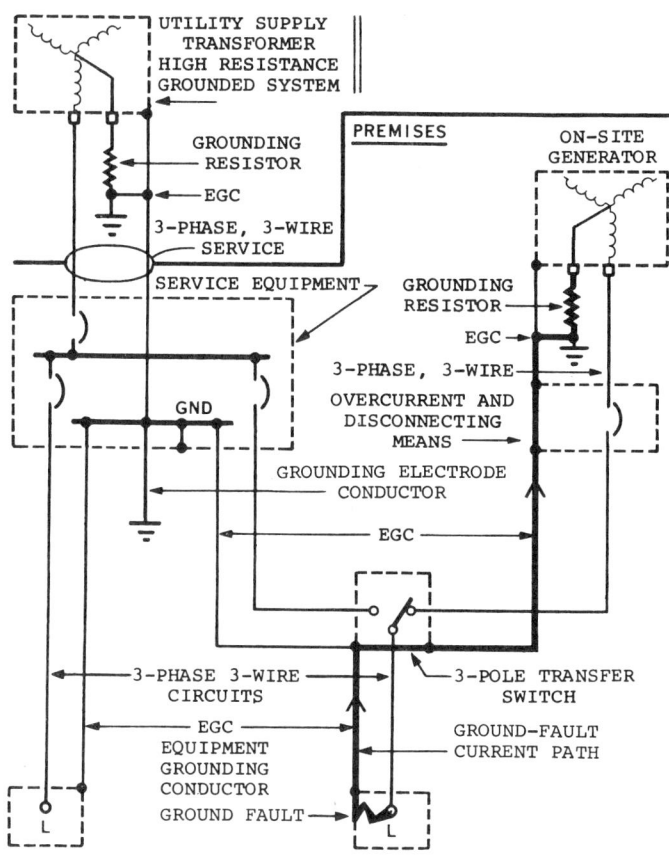

Fig 78
Ground-Fault Current Return Path to
High Resistance Grounded Alternate Supply

locations designated for operating the mobile equipment. If mobile generators are operated at locations where permanent grounding electrodes are not available, and the generator neutral is required to be grounded, plate electrodes should be considered because they can be installed quickly. Plate electrodes for system grounding should be located close to the generator and should be buried at least 1 ft below the surface of the earth. However, this approach should not detract from the desirability of connecting the generator neutral to a permanent grounding electrode.

Equipment grounding conductors must be provided with the mobile generator circuit conductors as illustrated in the diagrams in this section. If flexible cords or cables are used to connect mobile generators to emergency or standby power systems, each cord or cable must have an equipment grounding conductor.

Supplemental equipment bonding, which is discussed in 7.3, will reduce the risk of electrical shock for persons who contact the exposed noncurrent-carrying metal parts of the mobile equipment. Supplemental equipment bonding should be installed between the mobile generator frame and adjacent conductive surfaces such as structural steel, metal piping systems, and metal equipment enclosures. If the neutral of a mobile generator is not grounded, the generator frame should be connected to a grounding electrode, as illustrated in Fig 73, in addition to its connection to the equipment grounding conductor network.

7.12 Standards References

The following standards publications were used in preparing this section.

IEEE Std 141-1976, Recommended Practice for Electrical Power Distribution for Industrial Plants

IEEE Std 142-1972, Recommended Practice for Grounding of Industrial and Commercial Power Systems

NEMA No pB 1.2-1977, Application Guide for Ground Fault Protective Devices for Equipment.

NFPA No 70-1978, National Electrical Code

7.13 Bibliography

[1] BLOOMQUIST, W.C., OWEN, K. J., and GOOCH, R. L. High-Resistance Grounded Power Systems—Why Not. *IEEE Transactions on Industry Applications*, vol IA-12, Nov/Dec 1976, pp 574-579.

[2] CASTENSCHIOLD, R. Grounding of Alternate Power Sources. *Conference Record of the 1977 IEEE Industry Applications Society, Technical Conference*, paper 77CHl1246-8-IA, pp 67-72.

[3] CASTENSCHIOLD, R. Ground-Fault Protection of Electrical Systems With Emergency or Standby Power. *IEEE Transactions on Industry Applications*, vol IA-13, Nov/Dec 1977.

[4] WEST, R. B. Equipment Grounding for Reliable Ground Fault Protection in Electrical Systems Below 600 V. *IEEE Transactions on Industry Applications*, vol IA-10, Mar/Apr 1974, pp 175-189.

[5] ZIPSE, D. W. Multiple Neutral to Ground Connections. *Conference Record of the 1972 IEEE Industrial and Commercial Power Systems, Technical Conference*, paper 72CH0600-7-IA, pp 60-64.

[6] KAUFMANN, R. H. Some Fundamentals of Equipment-Grounding Circuit Design. *AIEE Transactions (Applications and Industry)* Pt II, vol 73, Nov, 1954, pp 227-232.

[7] WEST, R. B. Grounding for Emergency and Standby Power Systems. *IEEE Transactions on Industry Applications, vol IA-15, Mar/Apr 1979, pp 124-136.*

8. Industry Applications

8.1 Introduction

The information in this section is presented by the corresponding IEEE Industry Applications Society Committees.

8.2 Glass Industry

8.2.1 *Introduction.* The size of any emergency or standby power system is a matter of basic engineering economics in weighing system cost against production losses and their cost. The need is based on the type of production line and power system reliability.

The effects on product quality for various periods of time, that is, 1 min, 5 min, ½ h, 1 h, etc, must be taken into account in deciding what size standby system should be used.

Electrical dependence has been increasing with the advent of increased electric boosting, all-electric furnaces, electric forehearths, computer control, etc. It is necessary to develop a list of decreasing importance of various items critical to the glass production system to size a standby unit. The primary concern would be the safety of personnel, followed by the protection of the glass furnace.

8.2.2 *Applications.* For the melting department alone, consideration must be given to compressed air for oil atomization, perhaps pneumatic valve operators, etc. Combustion air is important, although furnaces may be operated on a natural draft basis through the opening of bulk heads. Block cooling fans or water pumps for water cooling systems may be extremely critical depending on the age and the condition of the furnace. Most controls may be bypassed for manual operation; however, such little power is required for many controls or readouts, such as temperature, auto reversal, etc, that they may be included.

Electric boosting is an important item. However, the percentage of the electrical input versus the total must be considered in sizing the standby unit.

Generally, voltage dips and infrequent, brief outages will not justify an emergency or standby power system cost since there is little if any effect on quality.

In the forming department, production interruption and the resulting cost of lost ware must be weighed against the cost of a standby system. Only a "bumpless" transfer system will prevent any ware loss. Therefore, typical outage time is again the criteria for determining standby power needs.

Similar analyses are required for all segments of the glass production line: annealing, finishing, packaging, etc.

In many cases, where a power outage may cause excessive losses to either production equipment or product, alternate energy, driving, or protective sources are part of the initial design.

Anticipated long term loss of power may be compensated for with a large portable generating unit tied directly into the plant distribution system at an appropriate point.

8.3 Rural Electric Power

8.3.1 *Introduction.* Electric power has become essential for agricultural production and loss of power can result in substantial financial losses, particularly when there are total confinement buildings, milking machines, mechanical feeders, or other such production equipment. Standby power is a form of insurance against power failure losses, and the cost of standby generation should be considered in relation to the possible monetary loss and inconvenience of a power failure.

8.3.2 *Full-Load Systems.* If it is desired to carry the full system load, the alternator must have sufficient capacity to carry the total emergency load as well as the ability to start the largest motor load on the system. Generally, automatic sequence starting must be used to connect individual loads one at a time.

8.3.3 *Part-Load Systems.* With a part-load system, only the most essential equipment is operated on the standby system. When a power failure occurs, all electrical equipment is turned off or disconnected. With the generator operating, essential equipment is turned on sequentially, starting with the heaviest loads until the capacity of the alternator is reached.

8.3.4 *Transfer Switch.* An essential part of a standby electric power system is a double-throw transfer switch to prevent the interconnection of the standby generator to the normal source of electric service. This requirement is necessary to conform with Article 750 of the National Electrical Code (NEC) and to prevent power from feeding back into the power supplier's lines and endangering the lives of linemen who may be working to restore power. The use of the transfer switch also prevents possible damage to the standby generator from interactions with the regular power service. The capacity of the double-throw switch must be matched to the rating of the conductors supplying the load from the normal electrical service and not to the rating of the standby source or the load supplied when the generator is in operation.

8.3.5 *Basic Standby Generator Types.* Standby generator units used for rural service are of two general types, tractor driven and engine driven. Tractor-driven units can be either stationary or portable and are driven from the power takeoff (PTO) shaft. The PTO speed changer must correspond to the proper PTO shaft speed (540 or 1000 r/min).

The horsepower rating of the tractor used to drive the PTO-driven alternator should be at least 2 hp for each kilowatt of electric power to be produced.

Engine-driven units may be either automatic or manual. For the average farm, a nonautomatic unit used with a power-off alarm is generally satisfactory and will be much less expensive. It is essential, however, that the power failure alarm monitor the critical load rather than the main service.

Automatic standby units are driven by

an engine that starts automatically in case of a power failure. The farm wiring system is transferred automatically to the standby unit when it is up to speed and ready for operation. The size of the unit required will be much larger than necessary unless some provision is made for starting up the loads in a sequential manner after a power interruption, starting with the largest motor or load.

8.3.6 *Sizing the Alternator.* The alternator selected must be chosen to match the power system with which it is to be used. For most applications, this will be a 60 Hz, single-phase unit for 120/240 V operation. The use of three-phase service is increasing, however, and some generators are available that may be used for either single or three phase. The capacity of the alternator must be sized to the largest load to be carried and sufficient capacity provided to start the largest motor on the system. Starting currents of large single-phase motors may run from three to six times the normal running current.

8.3.7 *Installation.* Proper installation of a standby alternator is essential for satisfactory operation. Details regarding proper methods and considerations can be found in the American Society of Agricultural Engineers (ASAE) Engineering Practice EP364 or the Electrical Generating Systems Marketing Association (EGSMA) publication, IMFS-1-1974.

In general, proper provisions must be made for suitable mounting, ventilation, fuel supply, exhaust, electrical connections, and accessibility for maintenance.

8.3.8 *Maintenance.* Maintenance instructions supplied by the manufacturer should be carefully followed to ensure maximum performance and reliability. Standby units should be kept in good running order at all times by exercising the unit at least once a month with a load connected. All generator openings should be covered by a ¼ in mesh screen to prevent damage by rats and mice.

8.3.9 *Accessory Equipment.* Meters should be provided so that the alternator can be operated at the proper frequency and voltage. Frequency meters or tachometers can be used to determine the correct generator speed. A voltmeter is also helpful in determining proper operating speed, and an ammeter is essential in preventing overloading of the alternator. An hour meter is recommended for engine-driven units to check on maintenance and servicing schedules. A pilot light connected to the power supplier's line is valuable in determining when normal power has failed and when it has been restored.

Alarms for indicating when power failures have occurred are essential in determining that power interruptions have occurred, particularly if the loss of power may cause a production loss. Many types of alarm systems are available for warning against almost any kind of undesirable condition. These range from a simple power-off alarm to sensors that will detect an interruption of air flow, temperature extremes, fire, smoke, or the malfunction of a piece of equipment. It is essential that the alarm detect power failure in the critical areas. A general power failure alarm may not provide adequate warning.

Alarm systems are not fail-safe. A convenient method of testing the alarm should therefore be provided.

8.4 Cement Industry

8.4.1 *Introduction.* Historically, cement plants have been departmental processes with material surge capacity between departments. However, modern, efficient plants approach a continuous process as economics indicate a greater utilization of waste heat, larger and fewer production units, less material storage,

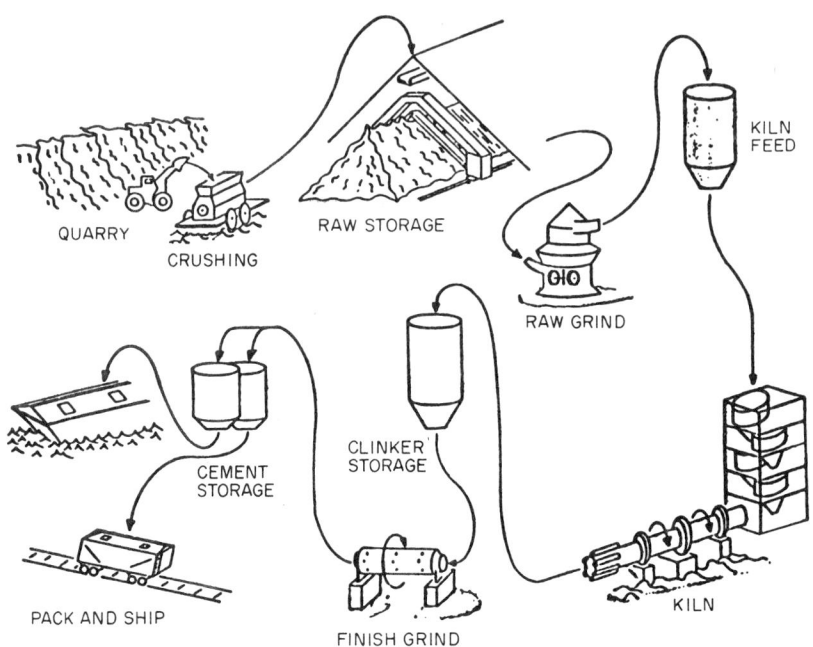

Fig 79
Flow Diagram for Typical Cement Plant

and centralized automatic control by computer. An electrical power interruption not only stops the process, it can wreak havoc with some capital intensive equipment. Thus, cement plant design includes standby power sources for both selected equipment and for an orderly shutdown of the computer.

Cement making consists essentially of quarrying limestone and clay or equivalent rocks, shells, sand, and waste materials. These materials are crushed and blended in chemical proportions and stockpiled for grinding. See Fig 79.

Raw grinding reduces the size of the material to 60 to 90 percent minus 200 mesh powder. Damp material will not grind easily, so the raw material is either slurried or dried for grinding. Present fuel cost requirement for subsequently evaporating water added to produce a slurry has essentially made the "wet process" obsolete. We now use the "dry process" as simultaneous dry and grind is far more efficient from an energy utilization standpoint.

The chemically balanced raw materials are heated in rotary kilns to about 2700°F (1500°C) where 20 to 25 percent melts, cement compounds are formed, and the red hot material rolls into pellets called clinker. These pellets are cooled to stabilize composition and to recover heat. Clinker is stored for subsequent use in the next step of the process.

Gypsum (3 to 5 percent) is added to the clinker and the combination is finely ground to 90 percent minus 325 mesh

(3000 to 5000 cm^2/g) to produce portland cement. Cement goes to storage for shipment as the finished product.

Dust collectors return dust to the process, or waste dust, as indicated by process chemistry.

8.4.2 *Immediate Need for Power.* The immediate need for power is one of safety, for both personnel and equipment. Partial loss of power initiates interlocks to automatically shut down related equipment, but some power is needed. These needs are handled separately as reliability of the power sources indicate.

(1) *Lighting.* Evacuation lighting, such as exit signs, exit lights, stairways, tunnels, and halls, are typically supplied by battery units which immediately and automatically switch to a charged nickel-cadmium battery for the period of the power outage. At least 1½ h of battery life is provided. Typical units are described in Section 4.

Also provided with battery backup lights are the electrical rooms, central control room, and warning lights. Since the power outage may extend beyond the battery capability, a tie to the emergency generator is recommended to sustain these lights, especially in the case of warning lights and the central control room.

Where auxiliary lighting for maintenance is desired, hand-carried or cart-mounted gasoline generators are sometimes used on extended power outages.

(2) *Valves, Dampers, Gates, Etc.* Valves, dampers, gates, etc, that must go closed or open on power failure for process control are handled separately. Fail-safe stored energy units are used to operate without external power automatically on power failure. Often, air operated units supplied from an accumulator are used to effect the fail-safe operation on loss of compressors, etc, when electrical power fails.

Where it is desired to monitor status of the process after a power failure, the emergency generator can power the central control panel and instrumentation. Thus, electrically operated instruments and instrument operators can be monitored and controlled after the power failure.

(3) *Water Pumps.* Water pumps for cooling water and for fire protection are important and a source of standby power must be provided. Usually, a separate diesel driven pump is provided to allow full standby service, in event of a pump, pump motor, or pump controller failure.

Automatic starting by battery or remote starting at central control is provided to quickly pump the needed water for cooling hot kiln and kiln fan bearings and for fire pumps. A wet quarry may also need some standby pumps for extended outages.

Sometimes a diesel generator supplies standby pumps. However, pumps are often remote from the generator and the separate diesel driven pump is preferred to minimize generator exposure to long lines and cost.

(4) *Passenger Elevators.* The diesel generator can also supply plant passenger elevators. A means of communications between the elevator and central control is desirable.

8.4.3 *Periodic Need for Power.* The periodic need for power is one of safety to equipment, primarily the kiln. Immediately upon power failure, fuel is stopped to the kiln. However, the kiln is hot and can be damaged by warping if it remains more than a few minutes without turning, especially when differentially cooled by rain. Some kilns have dedicated engines provided for this purpose. Also, cooling fans blowing air on kiln and clinker cooler TV cameras and clinker cooler bearings quickly need power to prevent overheating and subsequent damage. In wet process plants the agitators for slurry tanks should not be stationary for more than several hours or slurry may settle and jam the agitators.

In general, a diesel generator, either automatically started or remotely started from central control, provides the source of power to remotely operate these drives. See Fig 80. Some of the single motors are in the order of 100 hp and starting requirements for these squirrel cage induction motors usually determine the size and characteristics of the diesel generator. Typical sizes range between 100 and 600 kVA.

Diesel is mentioned. However, propane engines are also used for prime movers. Gasoline engines are not recommended because of hazards and evaporation which sometimes have left the user out of gas when he needed it.

Also, since the diesel engines could operate intermittently and for periods at low loads, an ambient lighting load of at least 25 percent of generator kilowatts is suggested for engine cleanliness.

Generators are usually 480 V, three-phase, 60 Hz, and feed selected motor control centers with transfer switches that automatically seek the source of power. Generators are intended to be isolated from utility power.

A specification for a cement plant emergency engine generator set appears at the end of this section. Typical units are described in Section 4.

A typical sequence appears as follows.

(1) Upon power failure, the engine generator is started. When up to voltage, the TV and cooler bearing fans are started and run while the generator is running. The kiln nose ring blower is started. These are fractional and integral horsepower motors.

(2) Next, the kiln emergency drive is started and the kiln rotated continuously

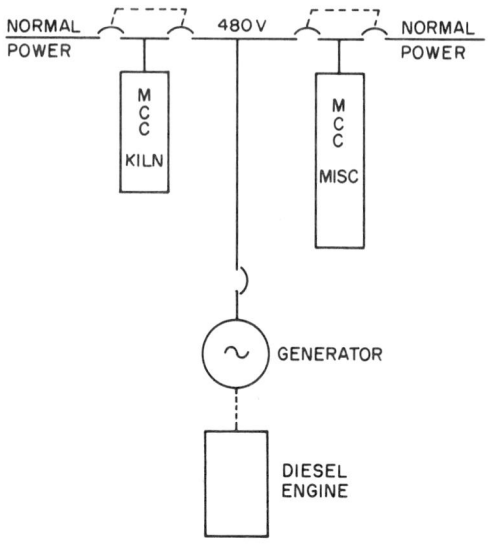

**Fig 80
Diesel Generator Emergency
Power for Cement Plant**

or one-quarter turn every 5 to 10 min. Continuous rotation could flood hot clinker into the clinker cooler and may damage the cooler. An exception is made during a rainstorm, when the kiln must be rotated continuously. Periods depend on conditions. These motors may be up to 100 hp each for two large kilns.

(3) Next, the clinker cooler quench fan is started as necessary, and runs as the air cooled hot clinker is discharged from the kiln.

(4) Other essential drives such as water pumps and slurry agitators are started. Chosen lights are transferred to the generator.

(5) After return to normal power, the generator is stopped and loads picked up at the discretion of the plant operator. On prolonged outages of several hours or more, the diesel generator may be stopped periodically and only restarted when needed.

To help ensure the diesel generator is available when needed, the unit is tested every week or two for proper operation. A careful operator may elect to have the generator set running during a thunderstorm in anticipation of a power outage during the worst time for a hot kiln. Operating history of the location and power system would be the guiding factor.

8.4.4 *Sustained Short Time Need for Power.* The sustained short time need for power is for the plant using a process control computer. This source can be for several seconds to allow for an orderly computer shutdown upon loss of power. Or, the source can be for a number of minutes to allow for the computer to stand by without loss of computer capabilities upon a power loss.

A flywheel motor generator set can provide a simple inertia driven "ride-through" system for the several seconds to allow an orderly computer shutdown. Also, the motor generator set helps isolate the computer from plant power system transients and electrical noise. This source is described in Section 4.

Typically, the computer is fed from a battery operated inverter. Batteries float on the power system. In event of loss of power from the inverter, power automatically switches to the plant power system. This source is also described in Section 4. A specification for a cement plant uninterruptible power system appears at the end of this section.

8.4.5 *Sample Specifications for Emergency Engine Generator Set*

(1.0) *Scope*

(1.1) This specification covers an emergency diesel generator set for emergency power outages. Set is to power squirrel cage induction motors and lights. Equipment anticipated to be started at full voltage by the generator is

1- _____ hp emergency kiln drive
1- _____ hp cooler quench fan
1-set of miscellaneous TV and bearings blowers estimated at _____ hp
1- _____ hp nose ring blower
Miscellaneous lighting and instrumentation power estimated at _____ kW.

(1.2) The specification is written for each engine generator set to be a complete entity to facilitate coordination required between engine generator set components. This coordination is included in this specification whether components are built by different divisions of one vendor, or by different vendors.

(2.0) *General*

(2.1) This specification covers electrical apparatus, completely coordinated for cement plant service. Each unit shall incorporate the features outlined below. Exceptions to the specifications shall be noted and priced separately.

(2.2) Ambient temperature to be 50°C maximum.

(2.3) Plant altitude is _____ m.

(2.4) Terminal blocks to show purchaser's wire numbers. Terminal blocks for purchaser's use or external sources of energy, or both, shall be disconnecting type. Internal wires are to be marked by wire numbers at both ends, with plastic sleeve markers.

(2.5) All equipment to have less than 90 dBA at 5 ft from equipment. Vendor to advise where noise levels will exceed this requirement.

(3.0) *Tests and Inspection*

(3.1) After assembly, NEMA standard tests shall be performed. At least, equipment shall be checked at twice rated voltage plus 1000 V.

(3.2) The electrical equipment included in this specification is to satisfy the requirements of NEMA, ANSI, OSHA, MSHA, and NEC.

(3.3) Supervisory services are part of this specification. Purchaser requires vendor to approve equipment installation, assist at startup, and train purchaser's personnel to assure proper operation of the equipment. This service is to be quoted separately.

(4.0) *Technical Specifications*

(4.1) *Mechanical Specifications*

(4.1.1) Engine generator set to be for mounting indoors in a cement plant, completely assembled on a bedplate (base), with vibration mounting included, and is also to include the following:

(a) 560 gallon oil tank for underground use, with tank oil level indicator.

(b) Fuel system (including 8 gallon day tank with brackets for wall mounting complete with float switch, auxiliary fuel pump, fuel lines between underground tank, fuel pump, day tank, and engine, if day tank is required for a fast start).

(c) Water jacket heater, complete with thermostat for 120 V, single-phase, 60 Hz, for cold weather starting. Power source by purchaser.

(d) Radiator and fan for 50 percent ethylene glycol solution.

(e) Duct between radiator and wall, and external wall gravity louver for engine cooling air, with flexible connection.

(f) Seamless flexible exhaust connection to muffler.

(g) Industrial muffler, with extension for building exit.

(4.1.2) Engine and engine generator with its control is to be suitable for operation in an ambient temperature of from − 20°F to + 120°F (−30°C to +50°C).

(4.1.3) Engine must be capable of starting immediately after shutdown, that is, engine *must not remain* idle for fuel pressures, etc, to subside to allow re-cranking and restart without overloading the auxiliaries or circuitry.

(4.1.4) Engine to be suitable for # 2 diesel fuel and also to be suitable to operate satisfactorily on # 2 fuel oil. Vendor is to recommend fuel and outline limitations.

(4.2) *Electrical Specifications*

(4.2.1) Diesel engine generator set to be _____ kW continuous, 0.8 power factor, _____ kVA diesel engine generator set, 1800 r/min, open dripproof, 480 V, three phase, 60 Hz, wye connected, to start and run squirrel cage induction motors, code F inrush, with static exciter and exciter regulator, oversized and fast enough to maintain no less than 85 percent volts (of 480 V) on full voltage starting of a _____ hp motor with _____ hp of induction motors running normally, and to be complete with the following:

(a) Control panel, completely assembled in a NEMA 12 enclosure for complete local control, with provision for remote control.

(b) 24 V electrical system, including starter, generator, batteries, battery rack, and 10 A regulated battery charger to operate from 120 V, single phase. A series parallel start-run is *not* acceptable.

(c) Provision for remote starting and stopping.

(d) Static exciter and voltage regulator to maintain ± 2 percent voltage regulation within the kilovoltampere rating of the generator, and to be large enough and fast enough to maintain voltage regulation outlined during starting.

(e) Governor to maintain frequency of generator to ± 3 percent with the kilovolt rating of generator.

(f) Main 600 V air circuit breaker or load break fused switch generator protection complete with shunt trip and 200 kA integrated circuit current limiting fuses.

(g) AC voltmeter, with provision for remote 150 V (for 600 V) voltmeter, with fused potential transformer.

(h) Ammeter and ammeter switch, complete with current transformers for three-phase amperes, and provision for remote metering of one-phase amperes with current transformer leads brought out to shorting terminal block on engine generator control, for reading on a 5 A, 4 VA basic instrument burden. Terminal block to be shipped shorted.

(i) Frequency meter and provisions for a remote meter.

(j) Running time meter.

(k) Charging rate ammeter.

(l) Oil pressure gage.

(m) Water temperature gage.

(n) Low oil pressure and high water temperature cutoff.

(o) Overspeed cutoff.

(p) Cycle cranking and cranking limiter.

(q) Oil level gage.

(r) Start-stop switch on control panel.

(s) Voltage adjusting rheostat.

8.4.6 *Sample Specifications for Uninterruptible Power Supply*

(1.0) *Scope*

(1.1) This specification covers an uninterruptible power supply (UPS) for ensuring continuous power to the plant process computer.

(1.2) This specification is for a complete package to facilitate coordination between components. This coordination is included in this specification whether components are built by different divisions of one vendor, or by different vendors.

(2.0) *General*

(2.1) This specification covers electrical apparatus, completely coordinated for cement plant service. Each unit shall incorporate the features outlined below. Exceptions to the specifications shall be noted and priced separately.

(2.2) Ambient temperature to be 40°C. Altitude is _____ m.

(2.3) Terminal blocks to show purchaser's wire numbers. Terminal blocks for purchaser's use or external sources of energy, or both, shall be disconnecting type. Internal wires are to be marked by wire numbers at both ends, with plastic sleeve markers.

(3.0) *Tests and Inspection*

(3.1) After assembly, NEMA standard tests shall be performed.

(3.2) The electrical equipment included in this specification is to satisfy the requirements of NEMA, ANSI, OSHA, MSHA, and NEC.

(3.3) Supervisory services are part of this specification. Purchaser requires vendor to approve equipment installation, assist at startup, and train purchaser's personnel to assure proper operation of the equipment. This service is to be quoted separately.

(4.0) *Technical*

(4.1) A static uninterruptible ac power system of the transfer type, consisting of a static inverter, battery charger, storage battery, and static transfer switch, as described in this specification, shall be provided for stand-by power application to the plant process computer. The computer requires _____ kVA at a constant voltage of 120 V ac, 60 Hz, single phase.

(4.2) The system shall be connected so that the preferred source of power is through the batteries, inverter, and static switch. If the battery power should fail, or drop below the preset transfer level, the load shall be automatically transferred to the alternate ac source. When the dc power returns to normal, the load shall be transferred automatically back to the preferred source after a 3 s delay. Prior to any transfer the two sources must be synchronized to achieve a bumpless transfer.

(4.3) A manual bypass switch shall be provided to isolate the static transfer switch and take it out of service without interruption to the load. With the static switch bypassed, either power source may be selected to supply the load. The switch is to be mounted in the inverter enclosure.

(4.4) An alarm contact shall be provided to indicate if the preferred source has failed or the alternate ac source is power to the computer.

(4.5) The system shall be such that a loss of the preferred power supply will result in a maximum power interruption to the computer of ¼ cycle.

(4.6) The system shall be capable of delivering a ——kVA output to the computer for ½ h. The batteries shall be nickel-cadmium, lead-calcium, or lead-antimony and come with a stand, connecting links, see-through cases and hydrometer. Specific gravity indicators to each cell should be priced separately.

(4.7) A two-position selector switch is also required for isolating the batteries and charger for maintenance. Mount selector switch on rectifier cabinet.

9. Case Histories

9.1 Introduction

Case histories of user needs for emergency or standby power systems are documented in this section. Preventive maintenance and full-load emergency operation simulation to test the equipment is required for reliable operation of an emergency or standby power system. Several examples follow.

9.2 Determining User Needs

9.2.1 *Earthquake Damage.* A survey of earthquake damage to buildings was conducted following tremors in San Fernando, California. 174 elevators failed because electric generators were made inoperative.

9.2.2 *Computer Power.* A major manufacturing concern installed a voltage monitor on the input system to a computer room late in 1973. The input voltage was normally 208 V ac, three phase, 60 Hz. In five weeks seven transients were recorded. Six of the transients were voltage drops of more than 10 percent, one was a voltage "spike" of more than 10 percent. Four of the transients exceeded 8 ms in length; three were immediately followed by computer malfunctions. Four of the transients occurred during a Sunday evening hour. The data collected showed that the voltage transients were troublesome to operations and justified a study to overcome the problem.

9.3 Failures Due to System Design Deficiencies

9.3.1 *Improperly Located Transfer Switch.* A transformer failure in an eastern hospital resulted in evacuation of the pediatrics, maternity, nursery, and medical-surgical floors because the emergency power system failed to take over. Later investigation revealed that the emergency power system never received a signal to start because the single automatic transfer switch was located on the primary side of the failed transformer. Good engineering practice calls for multiple transfer switches located close to the load.

9.3.2 *Insufficient Compressed Air for Starting.* An installation to prevent loss of power for more than a few minutes to a critical production process utilized compressed air for starting a diesel generator. Upon loss of normal power the standby power system started and operated for 15 min. At the end of the 15 min normal power returned and the standby unit was shut off. Immediately, the normal power failed again. The air compressor that supplied the air receiver for starting the generator set was not connected to the standby power system. Thus, the air receiver did not have enough compressed air to start the generator set the second time. The normal power source was out for several hours the second time, causing thousands of dollars of product to be scrapped because the standby power system could not be started.

9.4 Failures Caused By Lack of Maintenance

9.4.1 *Air Base Immobilized.* An air base was immobilized for over 3 h after a storm caused a power interruption on the utility lines and the off line standby diesel driven generator would not start. The utility company had to complete temporary repairs before the expensive standby equipment at the air base was operational.

9.4.2 *Drawbridge Inoperable.* Lightning caused an interruption on a utility line which immobilized a drawbridge in a partly closed position. In addition, the radio control for the ships entering the channel went out of service. The standby gasoline engine driving a generator failed to start because the gasoline had evaporated.

9.4.3 *Mechanical Stored Energy System.* A mechanical stored energy system was installed so that critical combustion controls for heating highly volatile solvents would ride through the momentary outage resulting from the utility switching from normal to emergency feeders. Although the equipment was designed and built by a reputable electrical equipment manufacturer, the first year's maintenance was so high and the interruptions caused by the system occurred so often that operating personnel installed jumpers to bypass the equipment. The utility supply was more reliable than the buffer equipment.

9.5 Misapplications of Emergency or Standby Power Systems

9.5.1 *Computer Power.* Illustrative of the kind of problem which can result when computer systems are operated directly from engine generator set power is an airport where diesel engine generators were installed to back up the flight control tower functions. When the utility supply failed during the great Northeast power blackout, these diesel engine generators were successfully started and alternating-current power from the sets was applied to the computer bus within a few minutes. But as a result of the short-duration loss of input alternating-current power and the voltage and frequency transients encountered in the changeover to diesel engine generator set power, the computer system suffered extensive damage which took several days to repair on an around-the-clock emergency basis during which time the control tower and flight operations had to be operated on a substantially reduced basis.

9.6 Successful Operations of Emergency or Standby Power Systems

9.6.1 *Bridges.* During a massive blackout in which a large utility lost its entire system, bridges and tunnels to an adjacent state supplied by another utility continued to operate because the alternate source was from the other utility which continued to operate. Bridges and tunnels supplied by two sources from the utility which lost its system were without power.

9.6.2 *UPS Survives Earthquake.* Two uninterruptible power systems, one located in Pasadena and one in Los Angeles, rode through the 1970 earthquake and worked perfectly when the utility company power was lost due to the quake. Both systems were installed to supply uninterruptible power to extensive computer systems.

Index

A

Acknowledgments, 208
Agricultural applications, 113, 194
Ampere-hour capacity, batteries, 122
Automatic transfer devices
 closed transition, 94
 features, 97
 protection, 95, 151, 157
 ratings, 95
 schemes, 94
Availability, definition, 19

B

Battery
 ampere-hour capacity, 122
 charging, 121, 124
 cost, 124
 maintenance, 149
 protection, 163
 type comparison, 124
Battery inverter systems, 126
Battery systems, 121
 application, 121
Boiler auxiliaries, 50
Brownout, 25
Bypass switches, 149

C

Case histories, 203
Cement industry applications, 195
Charging, battery, 121, 124
Circuit breakers, protection, 152, 154
Codes, state, 29
Cold start, 43
Commercial power, definition, 19
Communications systems, 78
Computer
 applications, 59
 auxiliaries, 68
 checklist for power requirements, 68
 cooling, 64
 definition, 19
 demand factor, 66
 effects on power supply, 65
 environment, 74
 harmonic distortion tolerances, 63
 inrush, 66
 need for emergency or standby power, 58
 power factor, 66
 power improvement techniques, 69
 power requirements, 60
 restarting, 57
 system classification, 57
 time shared, 58
 voltage imbalance tolerances, 64
 voltage tolerances, 61
Contract, utility, 21
Controller, automatic (process control), definition, 19
Conveyors, 47
Cost
 batteries, 124
 diesel generators, 85, 93
 gas turbine generators, 85, 107
 mechanical stored energy systems, 115, 116, 117
 mobile equipment, 112
 uninterruptible power supplies, 127, 132, 133, 135, 138, 139
 unit lighting equipment, 124

D

Data processing, definition, 19
Data processing equipment
 applications, 59
 auxiliaries, 68
 checklist for power requirements, 68
 cooling, 64
 demand factor, 67
 effects on power supply, 65
 environment, 74
 harmonic distortion tolerances, 63
 inrush, 66
 need for emergency or standby power, 58
 power factor, 66
 power improvement techniques, 69
 power requirements, 60
 restarting, 57
 system classification, 57
 time shared, 57
 voltage imbalance tolerances, 64
 voltage tolerances, 61
Data processor, definition, 19
Demand factor, data processing equipment, 67
Diesel driven generators, 26, 85
 advantages and disadvantages, 93
 air supply and exhaust requirements, 92
 automatic systems, 91
 controls, 87
 cost, 85, 93
 derating, 85
 fuel systems, 92
 maintenance, 103
 manual systems, 91
 motor starting considerations, 90
 noise, 92
 parallel operation, 88
 ratings, 90
 regulation, 92
 reliability, 91
 starting methods, 93
 transient loads, 91
 typical systems, 87
Disturbances, power, 21
Disturbances, voltage, 26
Dropout voltage (or current), definition, 19

E

Elevators, 45
Emergency power system, definition, 19
Engine driven generator, 26, 85
 advantages and disadvantages, 93
 air supply and exhaust requirements, 92
 automatic systems, 91
 controls, 87
 cost, 85
 derating, 86
 fuel systems, 92
 gasoline, 85
 maintenance, 144
 manual systems, 91
 motor starting considerations, 90
 multiple, 86
 noise, 92
 parallel operation, 88
 ratings, 90
 regulation, 92
 reliability, 91
 starting methods, 93
 transient loads, 91
 typical systems, 87
Environment, data processing equipment, 74
Equipment bonding, 170
Equipment grounding, 167, 169, 174
Escalators, 47
Evacuation lighting, 42

F

Fire protection, 56
 applications, 57

INDEX

feeder routing, 57
typical needs, 56
Firm power, definition, 19
Flywheel systems, 115
Forced outage, definition, 19
Frequency droop, definition, 20
Frequency regulation, definition, 20
Fuel systems
engine driven generators, 92
protection, 162

G

Gasoline engine generators, 85, 144
Gas turbine generators, 104
advantages and disadvantages, 93
cost, 93, 107
maintenance, 144
noise, 106
Generators
maintenance, 144
protection, 159
Generators, diesel driven, 85
advantages and disadvantages, 93
air supply and exhaust requirements, 92
automatic systems, 91
controls, 87
cost, 85, 93
derating, 86
fuel systems, 92
gasoline, 85
maintenance, 144
manual systems, 91
motor starting considerations, 90
multiple, 86
noise, 92
parallel operation, 88
ratings, 90
reliability, 91
starting methods, 93
transient loads, 91
typical systems, 87
Generators, gas engine, 85
Generators, gasoline engine, 85
Generators, gas turbine, 104
advantages and disadvantages, 93
cost, 93, 107
maintenance, 144
noise, 106
Glass industry applications, 193
Ground currents, 171
Grounding, 167
arrangements, 174
conductors, 173
equipment, 167
high resistance, 188
mobile equipment, 189
requirements, 172
system, 167

H

Harmonic content
definition, 20
tolerances, data processing equipment, 63
Health-care facilities, 74
Heating, process, 49

I

Industry applications, 193
Inverter, protection, 164
Inverter systems, battery, 126

L

Laws, state, 29
Life safety and life support systems, 74
Lighting
central battery systems, 125

cost, unit equipment, 124
evacuation, 42
for startup, 43
multiple power sources, 126
perimeter, 42
production, 42
recommended, 43
security, 42
supplemental, 43
system type selection, 125
unit equipment, 124
warning, 42
Load shedding, definition, 20

M

Maintenance
agricultural applications, 194
automatic transfer switches, 149
batteries, 148
diesel engines, 144
gas turbines, 145
generators, 146
mobile equipment, 112
uninterruptible power supplies, 147
Measurements, power disturbances, 23, 24
Mechanical stored energy systems, 114
cost, 115, 118, 119, 120
Mechanical utility systems, 47
Mobile equipment, 108
applications, 122
cost, 113
grounding, 189
maintenance, 112
rental, 112
Motor loads
starting considerations, 90
transferring, 99

N

Noise
electrical, 27
diesel driven generators, 92, 93
gas turbine generators, 93, 104

O

Off line, definition, 20
On line, definition, 20
Outage, forced, definition, 19
Outage, scheduled, definition, 20

P

Perimeter lighting, 42
Power
commercial, definition, 19
firm, definition, 19
utility, definition, 20
Power buffer performance, 119
Power disturbances
causes, 21, 23, 24
measurements, 23, 24
Power prime, definition, 20
Power factor, data processing equipment, 66
Power failure
definition, 20
duration tolerances, 28
examples, 53
table, 71
Power improvement techniques, 69, 71, 72
Power outage, definition, 20
Power, emergency systems, definition, 19
Power, prime, definition, 20
Power requirements, data processing equipment, 60
Power reserves, 53
Power, standby systems, definition, 20
Process heating, 49

Production, 42, 51
Protective equipment, 167
  causes of power disturbances, 21
Protection, 151
  batteries, 163
  circuit breakers, 152, 154
  fuel systems, 162
  generators, 159
  ground fault, 174, 175, 177, 178, 179
  uninterruptible power supplies, 163

R

Real time, definition, 20
Redundancy, definition, 20
Refrigeration, 50
Regulation, engine driven generators, 92
Regulations, state, 29
Rental, mobile equipment, 112
Rules, state, 29
Rural electric power applications, 113, 194

S

Scheduled outage, definition, 20
Security lighting, 42
Short circuit considerations, 151
Signal circuits, 80
Space conditioning, 54
  applications, 54
  codes and standards, 54
  data processing equipment, 64, 74
Standby power systems, definition, 20
Startup power, 43
State rules and regulations, 29
Steam turbine generators, 104
System grounding, 167

T

Thunderstorms, 21
Tolerance
  voltage, data processing equipment, 61

Tornadoes, 21
Transferring motor loads, 99
Transients
  causes, 24
  definition, 20
Turbine driven generators, 104, 105
  advantages and disadvantages, 93
  cost, 93, 107
  maintenance, 145
  mobile, 109
  noise, 105

U

Underground systems, 21
Uninterruptible power supplies
  battery selection, 137
  cost, 133, 134, 138, 139
  definition, 20
  nonredundant, 127, 135
  parallel, 132, 135
  protection, 163
  redundant, 132, 133
  reliability, 132
  rotating, 135
  with engine generator, 135
Utility power, definition, 20
Utility services, multiple, 94, 96
Utility, typical contract, 21

V

Voltage disturbances, 26
  causes, 21, 23, 24
  measurements, 23, 24
Voltage dropout, definition, 19
Voltage imbalance, 25
Voltage tolerance, 98
  data processing equipment, 61

W

Warning lights, 42

Acknowledgments

Appreciation is expressed to the following groups for their valuable assistance in preparing this document:

Canadian Standards Association (CSA), Standards Division, 178 Rexdale Boulevard, Rexdale, Ontario; Committee on Electrical Power Supply for Emergency Systems.
Computer and Business Equipment Manufacturers Association (CBE-MA), 1828 L Street NW, Washington, DC 20036.
Electrical Generating Systems Marketing Association (EGSMA), Tribune Tower, 435 N. Michigan, Chicago, IL 60611; Technical and Standards Committee.
Institute of Electrical and Electronics Engineers (IEEE); Power Engineering Society, Transmission and Distribution Committee, General Systems Subcommittee, Working Group on Service to Critical Loads.
Instrument Society of America (ISA), 400 Stanwix Street, Pittsburgh, PA 15222; Committee on Emergency Power Supplies SP54.
National Electrical Contractors Association (NECA), 7315 Wisconsin Avenue NW, Washington, DC 20014

Appreciation is expressed to the following companies and organizations for contributing the time of their employees to make possible the development of this text:

Automatic Switch Company
Bendy Engineering Co
Burroughs Corp
Corning Glass Works
El Paso Natural Gas Co
FMC Corporation
The Gates Rubber Co
Hensel Engineering
ICI Americas
Iowa State University
The Lima Electric Co
Monsanto Corporation
US Navy
Solar Turbines International
Stored Power Systems
Westinghouse Electric Corp